ADVANCING ROBUST MULTI-OBJECTIVE OPTIMISATION APPLIED TO COMPLEX MODEL-BASED WATER-RELATED PROBLEMS

Oscar O. Marquez-Calvo

ADVANCING ROBUST MULTI-OBJECTIVE OPTIMISATION APPLIED TO COMPLEX MODEL-BASED WATER-RELATED PROBLEMS

DISSERTATION

Submitted in fulfillment of the requirements of
the Board for Doctorates of Delft University of Technology
and
of the Academic Board of the IHE Delft
Institute for Water Education
for
the Degree of DOCTOR
to be defended in public on
Wednesday, 15 January 2020, at 15:00 hours
in Delft, the Netherlands

by

Oscar Osvaldo MARQUEZ-CALVO
Master of Science in Computer Systems Engineering,
Instituto Tecnologico y de Estudios Superiores de Monterrey (ITESM)
Master of Manufacturing Systems with major in Manufacturing Cells, ITESM
born in Oaxaca, Mexico

This dissertation has been approved by the
promotor: Prof. dr. D.P. Solomatine and
copromotor: Dr. J.L. Alfonso

Composition of the doctoral committee:

Rector Magnificus TU Delft	Chairman
Rector IHE Delft	Vice-Chairman
Prof. dr. D.P. Solomatine	IHE Delft / TU Delft, promotor
Dr. J.L. Alfonso	IHE Delft, copromotor

Independent members:

Prof.dr.ir. M. Kok	TU Delft
Prof.dr. Z. Kapelan	TU Delft
Prof. dr. M.J. Franca	IHE Delft / TU Delft
Prof. dr. P. Willems	KU Leuven
Prof.dr.ir. J.A. Roelvink	IHE Delft / TU Delft, reserve member

This research was funded by the National Council of Science and Technology of Mexico (CONACYT). Besides, it was conducted under the auspices of the SENSE Research School for Socio-Economic and Natural Sciences of the Environment. Cover photography by Gabriel Saldana - originally posted to Flickr as IMG_5031, CC BY 2.0, https://commons.wikimedia.org/w/index.php?curid=10791937, modified by adding the 3D graph.

Published by:
CRC Press/Balkema
Schipholweg 107C, 2316 XC, Leiden, the Netherlands
Pub.NL@taylorandfrancis.com
www.crcpress.com – www.taylorandfrancis.com

ISBN 978-0-367-46043-3

To my parents, my role models

To my siblings, my best and unconditional friends

To Sandra, who has always had a special place in my heart

ACKNOWLEDGMENTS

My sincere gratitude goes to my promotor Prof. Solomatine. Since the moment I met him for the first time, when I just arrived to IHE, his welcoming made me feel at the same time enthusiastic and worried because I knew that a steep way was coming. He also proposed the theme of this research which is very interesting and relevant. His critiques, inspiration, advice and constant feedback during the whole PhD shaped my research skills.

I also want to thank to my copromotor Dr. Leonardo Alfonso, who invited me to collaborate in some activities for the European Project KULTURisk for a term of one year. Later, we collaborated again in a research project related to water quality in water distribution systems, which has become one of the case studies. He then started to act as a co-supervisor and became my official copromotor. He was always finding time to guide me, advise me and help me in my research.

Additionally I want to thank Dr. Corzo Perez. He was the first to introduce me to the world of Hydroinformatics, to IHE Delft, and opened the opportunity to follow the PhD programme. I thank him for his enthusiasm and eagerness to provide help.

I especially thank the Mexican Government through the *Consejo Nacional de Ciencia y Tecnología* (CONACYT) which sponsored my PhD.

I would like to thank the staff of both institutes that hosted my PhD: IHE Delft and TU Delft. In particular I thank Jolanda, Anique, Gerda, Jos and Lilian for their prompt and kind help with my questions and my processes.

I also want to thank the people that made research projects with me. What began as collaborators and ended up in a friendship too. They are Alifta, Xu and Claudia.

Finally, I would like to thank my friends, who made very pleasant the moments of relaxation: Mario, Carlos, Laurens, Zahrah and Angeles.

SUMMARY

Optimisation has played an important role in the water sector. Since the 1960s (Karmeli et al. 1968; Schaake and Lai 1969) mathematical optimisation algorithms have been used. The tendency to apply optimisation methods is far from reaching a saturation point because every day it is touching more aspects such as policies, procedures, design, operation, etc. A number of these current trends are presented in Nicklow et al. (2009); Reed et al. (2013); Maier et al. (2014); Mala-Jetmarova et al. (2017).

In most applications, optimisation is carried out assuming deterministic variables. Only in the past 10 to 15 years, uncertainty started to be taken into account in the formulation of optimisation problems. Although this type of optimisation is generally named robust optimisation, the concept of robustness tends to differ from one study to the other. As a whole, robustness is determined depending on each case study or authors' preferences. Additionally, description or characterisation of the solutions' robustness is typically not explicit or not available.

Considering approaches reported in the literature to find robust solutions, we find it possible to group them in five categories, according to the way the optimisation problem is handled. The first category, the most common one, is the minimisation of the mean of the objective function rather than the minimisation of the objective function itself. For convenience this method can be named as OSOF (i.e. Optimisation by Smoothing the Objective Function). The second category considers minimisation of the mean and of the variance of the objective function. The third category adds an additional specific objective function to the set of objective functions of the original problem, which is related to robustness and it depends on the type of problem being optimised. The fourth category adds a constraint to the set of constraints of the original problem; the added constraint is related to robustness. Finally, the fifth category uses a technique of comparison of CDFs (i.e. Cumulative Distribution Function), which considers objective functions as random variables. It can be argued that all these approaches generate quite limited information about the propagation of the uncertainty from inputs or parameters to solutions.

Therefore, the following knowledge gaps in the current optimisation approaches to water related problems can be identified. First, most of the existing model-based optimisation algorithms used to solve water-related problems, in our opinion do not explicitly take uncertainty into account; that is, the resulting solutions are not robust and are sensitive to inaccuracies or uncertainty from different sources, including model inputs and model parameters. This may result in situations where the solutions, being optimal for one set of (deterministic) assumptions, are considerably sub optimal when moderate variations of these assumptions occur. Second, explicit estimates of the optimal solutions' uncertainty typically are not provided. Third, most of the robust algorithms are computationally intensive, so that executing them on one (personal) computer is very time consuming. Although this can be ameliorated by distributing the computational load among several computers, the parallelization of such algorithms is not always straightforward.

All these gaps prompt for advancing the algorithmic base of optimisation methods used in solving problems of water management. The proposed research aims to address the mentioned issues and focuses on developing and testing an algorithm for robust optimisation of multiple objectives. This developed algorithm is named Robust Optimisation and Probabilistic Analysis of Robustness (ROPAR).

The ROPAR algorithm is composed of four parts. The first part samples the input variables or parameters with uncertainty. The second part generates Pareto fronts by using any deterministic multi-objective optimisation algorithm. The third part is the visual or automated analysis of these Pareto fronts using probability density functions (i.e. PDFs), which are generated by selecting specific values of an objective function of interest and determining the distribution of solutions at this particular level. The fourth part is the selection of one or several robust solutions. This selection is carried out using robustness metrics reflecting various aspects of robustness presented in literature.

In this thesis, the ROPAR algorithm is tested initially on several cases, including a benchmark function and the problems of urban flood management and water distribution. For the benchmark function it is shown how the uncertainty is propagated to the solutions and how ROPAR allows to visualise the impact of this uncertainty. In this case study the parameter with uncertainty is a term added to the original formulation of the benchmark function.

ROPAR is also used to find robust solutions for the design of storm drainage systems. To this end, the objective functions considered are minimisation of construction costs and minimisation of flooding; the decision variables are the pipe diameters and the uncertain parameter considered is the rainfall. In order to study the advantages of robust optimisation, ROPAR was applied to one simple drainage system and to two complex storm drainage networks. Additionally, the OSOF method was also applied and the two methods compared. For the simple case, it is demonstrated that the solutions found using a deterministic optimisation are as robust as the ROPAR solutions. For the complex cases, it was found that ROPAR solutions are slightly better than the solutions found by OSOF.

Until this point all case studies have had just two objective functions and one source of uncertainty. Therefore, to challenge ROPAR, a problem with three objective functions and three uncertainty sources was also explored. The problem in this case was the design of a storm drainage system in combination with the implementation of stormwater Best Management Practices (i.e. BMP) in the same basin. In this case, the objective functions are minimisation of construction cost, minimisation of flooding, and maximisation of water infiltration to groundwater in different areas in the basin. Decision variables in this case are the pipe diameters and parameters related to BMPs, namely type, location and size of infrastructure to facilitate groundwater infiltration. The uncertain parameters under consideration are the rainfall, the age of the pipe and evolution of land-use in the basin.

ROPAR was also tested on a quite different problem – water quality problem in Water Distribution Networks (WDN). In this case study, the objective functions are minimisation of the valves number to be operated, and minimisation of water age in the network. The decision variables are the operational statuses of the valves (closed or not), and uncertain parameter is the water demand for each hour of a day (meaning that there are 24 uncertain parameters). Although a genetic algorithm (GA) was used to optimise the solutions in the previous cases, a faster optimisation algorithm suitable for the problem was developed. Once the efficiency of this new algorithm was tested and verified, it was used within the ROPAR algorithm to robustly optimise the solution to the water quality problem.

The conclusions of this research can be summarized in the following points. First, in principle, any kind of a deterministic multi-objective optimisation algorithm could be used within ROPAR. Second, ROPAR allows to estimate robustness of optimal solutions given uncertainty of inputs or parameters. Third, ROPAR is a method of general applicability. Fourth, ROPAR uses widely accepted robustness metrics and could use other robustness metrics. Fifth, ROPAR, when compared with the most common method used nowadays (OSOF), finds solutions with similar performance. Sixth, ROPAR is by design a computationally intensive method, but it can be straightforwardly parallelized, allowing for obtaining the results reasonably fast.

SAMENVATTING

Optimalisatie heeft een belangrijke rol gespeeld in de watersector. Sinds de jaren 1960 (Karmeli et al. 1968; Schaake and Lai 1969) worden wiskundige optimalisatie-algoritmen gebruikt. De neiging om optimalisatiemethoden toe te passen is verre van verzadigd omdat het elke dag meer aspecten raakt, zoals beleid, procedures, ontwerp, operatie, enz. Een aantal van deze huidige trends worden gepresenteerd in Nicklow et al. (2009); Reed et al. (2013); Maier et al. (2014); Mala-Jetmarova et al. (2017).

In de meeste toepassingen wordt optimalisatie uitgevoerd uitgaande van deterministische variabelen. Pas in de afgelopen 10 tot 15 jaar werd met onzekerheid rekening gehouden bij het formuleren van optimalisatieproblemen. Hoewel dit type optimalisatie doorgaans robuuste optimalisatie wordt genoemd, verschilt het concept van robuustheid van het ene onderzoek naar het andere. Als geheel wordt de robuustheid bepaald afhankelijk van elke case study of de voorkeuren van de auteurs. Bovendien is een beschrijving of karakterisering van de robuustheid van de oplossingen meestal niet expliciet of niet beschikbaar.

Gezien de in de literatuur gerapporteerde benaderingen om robuuste oplossingen te vinden, vinden we het mogelijk om ze in vijf categorieën te groeperen, afhankelijk van de manier waarop het optimalisatieprobleem wordt aangepakt. De eerste categorie, de meest voorkomende, is het minimaliseren van het gemiddelde van de objectieve functie in plaats van het minimaliseren van de objectieve functie zelf. Voor het gemak kan deze methode OSOF worden genoemd (d.w.z. optimalisatie door de objectieffunctie af te vlakken). De tweede categorie beschouwt minimalisatie van het gemiddelde en van de variantie van de objectieve functie. De derde categorie voegt een extra specifieke objectieve functie toe aan de set objectieve functies van het oorspronkelijke probleem, die verband houdt met robuustheid en afhankelijk is van het type probleem dat wordt geoptimaliseerd De vierde categorie voegt een beperking toe aan de verzameling beperkingen van het oorspronkelijke probleem; de toegevoegde beperking houdt verband met robuustheid. Ten slotte gebruikt de vijfde categorie een vergelijkingstechniek van CDF's (d.w.z. cumulatieve distributiefunctie), die objectieve functies als willekeurige variabelen beschouwt. Er kan worden betoogd dat al deze benaderingen vrij beperkte informatie genereren over de verspreiding van de onzekerheid van inputs of parameters naar oplossingen.

Daarom kunnen de volgende kennisleemten in de huidige optimalisatiebenaderingen voor water gerelateerde problemen worden geïdentificeerd. Ten eerste houden de meeste van de bestaande modelgebaseerde optimalisatie-algoritmen die worden gebruikt om water gerelateerde problemen op te lossen, naar onze mening geen expliciete rekening met onzekerheid. Dat wil zeggen dat de resulterende oplossingen niet robuust zijn en gevoelig zijn voor onnauwkeurigheden of onzekerheid uit verschillende bronnen, inclusief modelinvoer en modelparameters. Dit kan leiden tot situaties waarin de oplossingen, die optimaal zijn voor één set (deterministische) veronderstellingen, aanzienlijk suboptimaal zijn wanneer zich matige variaties van deze veronderstellingen voordoen. Ten tweede worden meestal geen expliciete schattingen van de onzekerheid van de optimale oplossingen gegeven. Ten derde zijn de meeste robuuste algoritmen rekenintensief, zodat het uitvoeren op één (personal) computer erg tijdrovend is. Hoewel dit kan worden verbeterd door de rekenbelasting over verschillende computers te verdelen, is de parallellisatie van dergelijke algoritmen niet altijd eenvoudig.

Al deze lacunes zijn aanleiding voor het bevorderen van de algoritmische basis van optimalisatiemethoden die worden gebruikt bij het oplossen van problemen met waterbeheer. Het voorgestelde onderzoek beoogt de genoemde problemen aan te pakken en richt zich op het ontwikkelen en testen van een algoritme voor robuuste optimalisatie van meerdere doelstellingen. Dit ontwikkelde algoritme heet robuuste optimalisatie en probabilistische analyse van robuustheid (ROPAR).

Het ROPAR-algoritme bestaat uit vier delen. Het eerste deel bemonstert de invoervariabelen of parameters met onzekerheid. Het tweede deel genereert Pareto-fronten met behulp van een deterministisch multi-objectief optimalisatie-algoritme. Het derde deel is de visuele of geautomatiseerde analyse van deze Pareto-fronten met behulp van waarschijnlijkheidsdichtheidsfuncties (d.w.z. PDF's), die worden gegenereerd door specifieke waarden van een van belang zijnde objectieve functie te selecteren en de verdeling van oplossingen op dit specifieke niveau te bepalen. Het vierde deel is de selectie van een of meerdere robuuste oplossingen. Deze selectie wordt uitgevoerd met behulp van robuustheidsmetingen die verschillende aspecten van robuustheid weerspiegelen die in de literatuur worden gepresenteerd.

In dit proefschrift wordt het ROPAR-algoritme in eerste instantie getest op verschillende gevallen, waaronder een benchmarkfunctie en de problemen van stedelijk overstromingsbeheer en waterdistributie. Voor de benchmarkfunctie wordt getoond hoe de onzekerheid wordt doorgegeven aan de oplossingen en hoe ROPAR het mogelijk maakt om de impact van deze onzekerheid te visualiseren. In deze case study is de parameter met onzekerheid een term die wordt toegevoegd aan de oorspronkelijke formulering van de benchmarkfunctie.

ROPAR wordt ook gebruikt om robuuste oplossingen te vinden voor het ontwerp van stormafvoersystemen. Daartoe zijn de beoogde objectieve functies minimalisering van bouwkosten en minimalisering van overstromingen; de beslissingsvariabelen zijn de pijpdiameters en de beschouwde onzekere parameter is de regenval. Om de voordelen van robuuste optimalisatie te bestuderen, werd ROPAR toegepast op één eenvoudig afwateringssysteem en op twee complexe stormafwateringsnetwerken. Bovendien werd de OSOF-methode ook toegepast en werden de twee methoden vergeleken. Voor het eenvoudige geval is aangetoond dat de oplossingen die zijn gevonden met een deterministische optimalisatie even robuust zijn als de ROPAR-oplossingen. Voor de complexe gevallen werd vastgesteld dat ROPAR-oplossingen iets beter zijn dan de oplossingen van OSOF.

Tot nu toe hadden alle case studies slechts twee objectieve functies en één bron van onzekerheid. Om ROPAR uit te dagen werd daarom ook een probleem met drie objectieve functies en drie onzekerheidsbronnen onderzocht. Het probleem in dit geval was het ontwerp van een afwateringssysteem voor stormen in combinatie met de implementatie van Best Management Practices voor regenwater (d.w.z. BMP) in hetzelfde bassin. In dit geval zijn de objectieve functies minimalisatie van bouwkosten, minimalisering van overstromingen en maximalisatie van waterinfiltratie naar grondwater in verschillende gebieden in het bassin. Beslissingsvariabelen in dit geval zijn de pijpdiameters en parameters met betrekking tot BMP's, namelijk type, locatie en grootte van infrastructuur om infiltratie van grondwater te vergemakkelijken. De onzekere parameters in kwestie zijn de regenval, de leeftijd van de buis en de evolutie van het landgebruik in het bekken.

ROPAR werd ook getest op een heel ander probleem - het probleem met de waterkwaliteit in waterdistributienetwerken (WDN). In deze case study zijn de objectieve functies minimalisatie van het aantal te bedienen kleppen en minimalisatie van de waterleeftijd in het netwerk. De beslissingsvariabelen zijn de operationele statussen van de kleppen (gesloten of niet), en onzekere parameter is de waterbehoefte voor elk uur van een dag (wat betekent dat er 24 onzekere parameters zijn). Hoewel in de vorige gevallen een genetisch algoritme (GA) werd gebruikt om de oplossingen te optimaliseren, werd een sneller optimalisatie-algoritme ontwikkeld dat geschikt was voor het probleem. Nadat de efficiëntie van dit nieuwe algoritme was getest en geverifieerd, werd het binnen het ROPAR-algoritme gebruikt om de oplossing voor het waterkwaliteitsprobleem robuust te optimaliseren.

De conclusies van dit onderzoek kunnen op de volgende punten worden samengevat. Ten eerste zou in principe elke vorm van een deterministisch multi-objectief optimalisatie-algoritme kunnen worden gebruikt binnen ROPAR. Ten tweede maakt ROPAR het mogelijk om de robuustheid van optimale oplossingen te schatten, gezien de onzekerheid van inputs of parameters. Ten derde is ROPAR een methode voor algemene toepasbaarheid. Ten vierde gebruikt ROPAR algemeen aanvaarde robuustheidsmetingen en zou het andere robuustheidsmetingen kunnen gebruiken. Ten vijfde vindt ROPAR, vergeleken met de tegenwoordig meest gebruikte methode (OSOF), oplossingen met vergelijkbare prestaties. Ten zesde is ROPAR een berekening intensieve methode, maar deze kan eenvoudig worden geparallelliseerd, waardoor de resultaten redelijk snel kunnen worden verkregen.

CONTENTS

Contents

1 INTRODUCTION

This chapter presents the background of the problem, the motivation for this research, and the objectives. It also describes the structure of the thesis.

1.1 BACKGROUND

Optimisation methods require a diverse set of mathematical and information technology tools. In particular for water-related problems, the use of mathematical models and simulations is critical. In addition, the use of optimisation techniques allows for identifying the best management options, and probabilistic approaches help to evaluate and handle uncertainty. In the following lines background on these concepts are presented.

1.1.1 Models and simulation

In general, a solution to any engineering problem can be approached by the combination of at least three aspects, namely (1) applied theory and reasoning, (2) experience, and (3) model experimentation. Sometimes the first two aspects are not enough, either because the problem is too complex to be solved theoretically or because there is no experience yet in solving the problem. In those cases, experimenting with a model is a welcomed alternative to add to solve the problem.

Models are mathematical tools that simplify reality in order to observe and understand how a complex system work. They allow to simulate the response of the system under different situations and inputs. Three model categories can be recognised (Novak et al. 2010), namely direct, semi-direct and indirect. The direct model is a reproduction of the real system to a smaller scale, the semi-direct model is an analogy to the real situation, and the indirect model is a representation of reality that uses theoretical analysis that can be mathematical, computational, numerical, or based on data.

In this thesis, the experiments are carried out using computational models built (instantiated) using computer-based modelling systems. The modelling systems used in this study are SWMM (Rossman 2010) for the case of modelling urban drainage systems, and EPANET (Rossman 1999) for the case of modelling drinking water distribution networks.

1.1.2 Uncertainty

Classical engineering problems often used to make a very strong assumption that the universe was unchanging, deterministic and to some extent predictable (Serrano 2011). Such simplifications allowed for a more easy understanding of the physical and chemical laws that govern the world. However, in real life there is variability that often stems from the unpredictable behaviour of nature. These unpredictable events are present in most engineering fields and natural processes, for example: the size and path of a storm, the intensity and behaviour of an earthquake, the number of defective products in a manufacturing line, the time of the next failure in a machine, the path followed in the subsurface by a spilled pollutant on the ground, etc. Among other reasons, we are uncertain about these events because we currently have incomplete or low quality information, or lack of knowledge.

Further, engineers often work not only with incomplete information, but also with incorrect information. This incorrect information could be due to human error; or it could be due to incorrect reading of the variable result of failure, miscalibration, or lack of resolution of the measuring device, and all these aspects increase uncertainty.

So, how can uncertainty be described in general terms? Zimmermann (2000) defines uncertainty as a process which cannot be appropriately anticipated neither deterministically nor numerically, due to the lack of enough information. Uncertainties can be characterised as epistemic and aleatory (Sullivan 2015). Aleatory uncertainties are those variations that are random by nature, for example, the path followed by a particle in a turbulent flow. Epistemic uncertainties, on the contrary, are the variations that cannot be explained because of lack of knowledge about the phenomenon, for example the geometrical tolerances in a structure produced by the manufacturing process, or structural changes of pipes in drainage systems. It should be mentioned that quite often it is difficult to attribute uncertainties to any of these types: e.g. uncertainty in rainfall has both aleatory and epistemic characteristics.

Particularly for the case of models, the epistemic uncertainty can be divided in two groups: model-form uncertainty and parametric uncertainty (Sullivan 2015). Model-form uncertainty is related to the lack of accuracy of the model to simulate the response of the real system. Parametric uncertainty is the uncertainty associated with various parameters associated with the model. It is often reasonable to further distinguish uncertainty in model parameters (e.g. roughness in drainage pipes), and in model inputs (e.g. rainfall). In this thesis focus is given to parametric and input uncertainty.

1.1.3 Optimisation and robust optimisation

Optimisation can be defined as the process of finding the best use of certain resources in order to reach a goal (or a set of goals) within certain constraints. From this definition, the three main elements that define an optimisation problem can be identified: first, the objective function (or objective functions) defined by the goal (or goals); second, the decision variables that specify how the resources are used; and third, the constraints specifying the limits in which the decision variables can be manipulated, since they are mainly associated to the availability of resources and conditions of their use.

In model-based optimisation, a model is used to evaluate the objective functions that are the result of a given set of decision variables that comply with certain restrictions. In this respect, this research is framed within a multi-objective optimisation, which is applied to complex water-related problems with the help of computational models. More specifically, a water-related problem (WRP), e.g. the design, management and/or operation of a water system, is simulated by a computational model, which is used to determine the optimal value of a set (vector) of decision variables. This process is guided by the iterative evaluation of the objective functions.

The traditional way of solving engineering problems, including WRPs, generally follows a deterministic approach. That is, engineers use to solve these problems by calculating design values and somehow tackling uncertainty by adopting extra measures in their design, such as the application of safety factors, in order to reduce the risks of failure. In risk management the presence of uncertainties is very well recognized, e.g. engineering probabilistic design is widely used. However, full consideration of all sources of uncertainty associated with the problems, and studying the ways of how such uncertainties actually influence those solutions (i.e. how uncertainties propagate through the optimisation process to solutions) still needs development of mathematical and algorithmic apparatus.

During the last decades, this problem has been recognized more and more, and one can see an increasing interest in developing methods to account, measure and visualise uncertainty. This viewpoint has also been boosted by the computing power (e.g., parallel and cloud computing) that has enabled the technological possibility to assess in a broader way the impact of a range of scenarios in a system, and in a broader range of systems, particularly hydrologic and hydraulic systems – see, e.g., Pappenberger and Beven (2006).

The concept of taking into account uncertainty in the optimisation process is often named robust optimisation. It aims to find a robust solution (i.e. a configuration of the system to be optimised). In this work we adopt the following definition.

> *A system is called robust if its performance, measured by objective functions, remains near the optimum value despite undergoing uncertain conditions.*

A "solution" is a set of decision variables (parameters) which can be identified by solving an optimisation problem, and which uniquely defines the 'system'. Robust optimisation is a procedure which leads to finding such robust solutions. More detailed definitions of robustness and its various features, and the corresponding mathematical definitions, are given in Chapter 3.

A number of methods for robust optimisation have been developed. The so-called 'stochastic programming' (Shapiro et al. 2009) offers a number of approaches to deal with the optimisation problems in the presence of uncertainties. In most problem settings, stochastic programming is in fact optimisation of the expected mean of the objective (or the sample average over a set of realizations generated using Monte Carlo simulation). Examples on the use of stochastic programming (stochastic linear programming and stochastic dynamic programming) in water related problems are given, e.g. in Loucks and Van Beek (2017). This approach offers a number of useful ideas and techniques and in some cases may work well, but it lacks the means of testing if indeed the generated solution(s) are robust against particular uncertainties. Another limitation is that it does not explicitly deal with indicators of robustness.

Water resources practitioners currently make little use of robust optimisation techniques (Basdekas 2014). This is due, among others, to the following factors: the complexity of relevant algorithms, the substantial computing power required for these algorithms, lack of readily available software, and perhaps the lack of skills to apply the novel optimisation and data processing techniques. These are the points addressed by this thesis.

It is worth mentioning that the method explored is computationally intensive, as any method using Monte Carlo analysis is, since the computational model is run many times. A possible way to reduce complexity is to control (decrease) the number of Monte Carlo runs. One may also avoid the use of Monte Carlo analysis, and to use other techniques, e.g. based on possibilistic uncertainties (fuzzy logic and evidence theory).

In the next section examples of uncertainty in water problems will be presented to show how important the problem is.

1.1.4 Examples of uncertainty in water related problems

Every system has associated uncertainties that must be identified to solve water related problems. In this section, four water system examples and their associated problems are presented, and some of the uncertain variables are identified. The water systems under consideration are urban drainage systems, water distribution systems and multi-purpose water reservoir systems. Additionally, infrastructure for flood protection is also considered.

Drainage systems consist of a set of infrastructural elements (e.g., pipes, canals and hydraulic structures) installed in urban or rural areas to drain the excess runoff water produced by rainfall and that can lead to flooding. In combined systems, this infrastructure also collects the wastewater produced in the city and transports it to treatment plants or to receiving water bodies. These water systems have problems associated to planning and management. In particular, the problem under consideration is the optimal design of these systems, which consists in determining the appropriate diameter of the pipes under some restrictions associated to costs, slopes, and limiting velocities. In this case, uncertain variables are the estimation of rainfall extremes (Arnbjerg-Nielsen et al. 2013), the evolution of land use in a subcatchment leading to changes in runoff patterns, the changes of pipe roughness due to aging, reduction of effective diameter due to accumulation of sediments, etc. To determine the capacity of channels/pipes in a rural drainage system, uncertain variables include the roughness coefficient of the canals and the pervious area depression storage. To implement Best Management Practices which include, among others, bio-retention cells, porous pavement, vegetative swales, and green roofs, the uncertainties are the conductivity slope, the surface roughness, the vegetation volume, etc. Although all the uncertainties just mentioned belong to flow models, other models, for example water quality models, can have many more uncertainties (Willems 2008).

Water distribution systems consist of sets of infrastructural elements (pipes, storage tanks, pumping stations, etc.), to safely and reliably transport and distribute, via a network of pipes, drinking water from a treatment plant to final consumers. Problems related to these systems are associated with planning (e.g., expansion, pipe rehabilitation, design) and management (e.g., operational status of elements according to specific needs). In this thesis we mainly concentrate on the design of operational strategies to control water quality, which contain uncertain parameters. For example, changes in the distribution of the population and land use over time, quantification of the user demands, leakages due to pipe failures, etc.

Reservoir systems are structures to store water for various purposes. The associated problems are also related to planning and management. Design variables are typically volume of storage, minimum and maximum water levels and minimum and maximum capacity of flow release and hydropower production; additional variables can be listed for the case of reservoir operation. Some of the uncertain variables to take into account in both problems are the amount of rainfall (Milly et al. 2008) and evaporation, the demand of electricity to be produced, and the demand of water for different water users such as ecology, agriculture, urban areas, etc.

Flood protection is typically associated with structural measures (dams, levees, retention basins, contingency basins, polders, etc.) to reduce flood risk, and non-structural measures such as early warning systems and risk communication. Design of structures is influenced by many uncertainties, for example the sea level increase, changes in economic value of the areas that to be flooded (Tsimopoulou et al. 2015), as well as climate change, tectonic subsidence, assumptions in the design of the infrastructure, price of sand, oil, etc. (Stijnen et al. 2014),

In all the problems mentioned, the design (or operation) exercise implies an optimisation procedure. The fact of not taking into account the corresponding uncertainties can lead to systems that underperform for the real working conditions which are likely to differ from those accounted for during the design (or operation) process, or systems which are 'over-engineered' and hence unreasonably expensive in implementation.

1.2 MOTIVATION

The motivation of this thesis is based on the fact that uncertainty in the optimisation exercise could be treated in a more consistent way. In this section a review of literature dealing with optimisation is shown. A number of comprehensive reviews on this subject worth reading are Nicklow et al. (2009), Reed et al. (2013), Maier et al. (2014) and Mala-Jetmarova et al. (2017).

Design and rehabilitation of storm drainage systems are classical problems where optimisation techniques are used. Guo et al. (2008) provide a complete review and some of the related works are mentioned further. For example, Barreto et al. (2009) consider optimisation of pipe diameters to minimise rehabilitation costs as well as flood damages. Velez Quintero (2012) finds the optimum pipe diameters and parameters of water treatment such as storage capacity and pumping flow to minimise floods and costs. Karovic and Mays (2014) minimise cost of the network through optimising pipe sizes and their corresponding slopes. Cozzolino et al. (2015) minimise construction cost of a rural network by optimising the geometric characteristics of open channels. Sebti et al. (2016) propose a way to find optimum designs for the restructuring of an urban network determining the most adequate types of Best Management Practices to minimise both runoff and modification cost. Steele et al. (2016) minimise construction costs by optimising the layout of the network. Yazdi et al. (2016) consider the problem of minimising both rehabilitation cost and flooding overflow volume by optimising pipe size, and the authors analyse the efficiency of several optimisation methods.

However, in these works there is an important aspect of practical optimisation which is not always taken into account, which is considering various types of uncertainty that may influence the choice of optimal solutions. In this regard, various authors frequently use the term 'robustness' with definitions, when provided, that differ from paper to paper. In relation to design of drainage systems, this aspect has been addressed in some of the research of the last decade. Maharjan et al. (2008) propose a method to introduce Best Management Practices deferred through the life span of the network to minimise its overall cost and also to adapt to the uncertain conditions of the future, namely land use, demography, and rainfall. The robustness is implicit in the solution because the optimum solution integrates the uncertainty of future conditions. Zeferino et al. (2012) analyse three alternative formulations of the objective function measuring the robustness of setting up and operating a wastewater system with uncertainties in the river flow. Andino-Santizo (2012) optimises the design of a drainage network using rNSGA-II taking into account uncertainty in the population growth. Kang and Lansey (2012) deal with the design of water and wastewater infrastructure through the minimisation of both the mean and standard deviation of its cost which include regret costs, besides the setup and operation costs. This type of approach is named 'smoothing of the objective function' or simply 'smoothing' in the rest of the document. The uncertainty considered in that paper is determined by five scenarios of water demand and wastewater production. Vojinovic et al. (2014) apply two methods (smoothing and rNSGA-II) to find the robust optimum rehabilitation/design of an urban network with respect to the diameters of the pipes. The uncertainties considered are rainfall, land use, demography and aging of pipes. Those authors define the 0% robustness when the original network is evaluated with the worst case (largest value) of the uncertain parameters, the 100% robustness – when there is zero damage in a design undergoing the worst case, and any other intermediate value of robustness is estimated by interpolation of these two extremes.

Further, Yazdi et al. (2014) optimise rehabilitation designs varying the pipe diameters, taking into account uncertainty of rainfall, modelled as a joint probability distribution density of its duration and intensity. The objective functions considered are the expected overflow of the network and the rehabilitation cost by averaging the objective functions (i.e. their smoothing). The robustness of every solution of the Pareto front is specified using its confidence interval. Kebede (2014) optimises the diameter of the pipes of a stormwater drainage system assuming the uncertainty in both Manning's roughness and rainfall using 'smoothed' objective function (damage cost) and the already mentioned rNSGA-II. Toloh (2014) and Martinez-Cano et al. (2014) optimise the resilience of urban drainage considering uncertainty in the rainfall (by resilience understanding the flooding cost multiplied by the probability of the return period occurrence), and applying for optimisation a standard genetic algorithm. Galindo-Calderon et al. (2015) optimise type and location of the Best Management Practices to minimise investment cost and peak runoff. Land use is the uncertain parameter considered. Using the objective function smoothing, robustness is calculated as the average of the ratio of the expected runoff (derived by considering samples of every possible land use scenario) to the maximum runoff of the base scenario (doing nothing).

It is worth noting the relevant contributions by researchers at the University of Exeter to the field of optimisation of urban networks. They mostly relate to water distribution systems (WDS). Babayan et al. (2004) optimise the resilience of the WDS by using the multi-objective optimiser NSGA-II (Deb et al. 2002) where in every evaluation of the objective function parameters with uncertainty are used to identify the critical nodes and significant variables. Savic (2005; 2006) finds the robust optimum design of WDS, and mentions that the same general methodology may be applied to find the robust optimised design of a drainage system. In the mentioned WDS case, the objective functions are the minimisation of the cost as well as the maximisation of the robustness, and the decision variables are the diameters of the pipes, and the considered parameter with uncertainty is the user demand. The robustness is defined as the percentage of the nodes that meet the minimum head requirement across the whole network. This kind of percentage is also generically known in engineering as *reliability* (Loucks and Van Beek 2017; Jin 2019), i.e. *reliability* is understood as a probability of system failure under certain conditions.

As a matter of fact, this notion of *reliability* is often used in Water sciences interchangeably as robustness. For example, Kapelan et al. (2005) and Kapelan et al. (2006) present the details of the methods to design of water distribution systems under uncertainty, employing rNSGA-II. These papers have considerably influenced the water science community to think about the optimisation problems in the presence of uncertainty.

In carrying out this literature review, the aim was to analyse various ways of robustness definitions, its analysis, and approaches to optimisation that would take uncertainty into account. It has been found that in most cases uncertainty (expressed probabilistically) is not explicitly propagated to the robust solution allowing for its analysis, but the most widely used approach to encounter for uncertainty is (artificially) making the objective function (OF) more 'robust'. This is achieved by 'smoothing' it in the neighbourhood of a given point applying some kind of a filter, e.g. by integrating or averaging across a number of points in a proximity of a given point, thus making OF less sensitive to variation in parameters or inputs. With this robust variant of OF, the standard optimisation algorithms are employed, typically, various versions of randomized search, like genetic algorithms. Such smoothing is, for example, an integral part of the so-called 'noisy' genetic algorithms (GAs) (which actually optimise the smoothed OF instead of the original one). In this class of algorithms OF at each point (chromosome) is assessed by averaging OF for several neighbouring points. Note however, that the resulting smoothing cannot be formally quantified since it is not based on mathematically formal statistical properties of the assumed uncertainty or those of the samples generated by GA. A variant of such algorithm for multi-objective optimisation, named rNSGA-II, and based on the popular NSGA-II algorithm of Deb, was presented by Kapelan et al. (2005)

and Kapelan et al. (2006). The approach to robust optimisation using smoothing of OF, will be further called 'Optimisation by Smoothing the Objective Function' (OSOF).

It can be also said that in OSOF (and in other methods to find robust solutions of multiple objectives), the uncertainty is, in fact, hidden in the identified optimum solutions, and for that reason propagation of uncertainty from inputs or parameters to solutions is not explicit and cannot be directly estimated or analysed. To work around this situation, some authors (Babayan et al. 2004; Kapelan et al. 2006; Erfani and Utyuzhnikov 2012; Zeferino et al. 2012; Toloh 2014; Marchi et al. 2016; Roach et al. 2016) have included an additional objective function that is measuring in some respect the robustness of the solution (for example for the case of WDS such function may measure resilience of the network). This is indeed a valid approach, but it also has certain drawbacks: it is problem-dependent, does not provide universal (probabilistic) instruments and metrics for explicit analysis of robustness and increases the dimension of the optimisation problem - which may make it more difficult to visualise Pareto sets even for two dimensional problems and thus reduces ability of decision makers to choose the best solution.

In summary, the analysis of literature and previous experiences in urban network optimisation prompts for developing a new optimisation method able to account for uncertainty in a more consistent way.

1.3 RESEARCH QUESTIONS

The literature review allowed to identify the specific research questions addressed in this study.

RQ1. What algorithm of optimisation of multiple objectives could be used as a 'deterministic optimisation engine' in the robust optimisation framework?

RQ2. How to estimate robustness of optimal solutions given uncertainty of inputs or parameters?

RQ3. What considerations must be taken into account by a generic framework to find robust solutions?

RQ4. How can the definition of robustness be extended to problems with multiple objectives?

RQ5. How does the proposed algorithm compare with the existing approaches to robust optimisation, and what is its computational complexity?

RQ6. Can the proposed framework be used for problems of different contexts and settings?

RQ7. What are the ways to ensure reasonable efficiency in obtaining robust solutions?

1.4 OBJECTIVES

Main objective

To review, improve and develop new algorithms capable of finding robust optimal solutions to complex problems in water management with multiple objectives, taking into account data uncertainty.

Specific objectives

These objectives are stated in order to answer the research questions. Figure 1 shows how these objectives are mapped to the thesis structure.

1. To review algorithms for optimisation of multiple objectives.

2. To develop a method to estimate robustness of optimal solutions given uncertainty of inputs or parameters.

3. To develop a framework of general applicability for robust optimisation of multiple objectives.

4. To propose an adequate and extended definition of robust solutions in the context of multi-objective optimisation.

5. To compare the developed method with current approaches of robust optimisation of multiple objectives.

6. To test the developed approach on benchmark functions (i.e. those used to test algorithms of optimisation of multiple objectives) and in real life cases related to urban drainage as well as water distribution networks.

7. To explore the ways of increasing efficiency of the proposed method for problems with multiple objectives.

1.5 INNOVATION, PRACTICAL VALUE AND SOCIAL RELEVANCE

The novelty of this research lies in several aspects.

1. The uncertainty of the parameters is explicitly propagated to the set of potential solutions which can be visually and mathematically analysed.

2. The method favourably compares with a wide class of methods using optimisation of the smoothed objective function.

3. The method offers the possibility of analysing the robustness of the solutions using not one but several dimensions of robustness.

4. The proposed approach has general applicability and can be applied to the optimisation of a wide variety of problems, where the uncertainty of input or parameters is represented probabilistically.

5. The proposed method is computationally intensive but it can be straightforwardly parallelized.

The practical value and social relevance of this thesis relies on the potential impact of robustly optimised systems, which will be able, on average, to perform well under different uncertain conditions. Specifically, the robust solutions of the cases shown in chapters 4, 5 and 6, which are about combined drainage systems, could improve the situations of minimising floods and therefore saving lives and reducing damage to property. Similarly, the robust solution of case study of Chapter 6 could maximise the infiltration to groundwater and minimise the contamination of water bodies by minimising the pouring of untreated drainage water under rainy conditions, impacting water scarcity and water pollution. Finally, the robust solutions of the cases of Chapter 7 regarding water distribution systems could improve the delivery of healthy drinkable water to the population.

To conclude, the framework developed in this thesis could allow engineers and managers make more informed decisions. Other kinds of problems, different from the ones solved in this thesis, could also be addressed.

1.6 THESIS STRUCTURE

The thesis structure is schematised in Figure 1.

Chapter 1 introduces the theme of the thesis as already explained.

Chapter 2 is devoted to revise the state of the art of three pillars of this research: first, algorithms of multi-objective optimisation; second, approaches of robust multi-objective optimisation used in water sciences; and third, methods of robust multi-objective optimisation in general. Additionally, the knowledge gaps are pointed out, which lead to present the research questions.

Chapter 3 presents the methodology. It begins by illustrating the components of the framework, followed by detailed explanation of every component. It also presents the plan of experiments to test the proposed approach named Robust Optimisation and Probabilistic Analysis of Robustness, ROPAR. An illustrative example (a benchmark function widely used in optimisation) is used to demonstrate the essence of the proposed approach. Finally, analysis of the computational efficiency of this approach is presented.

11

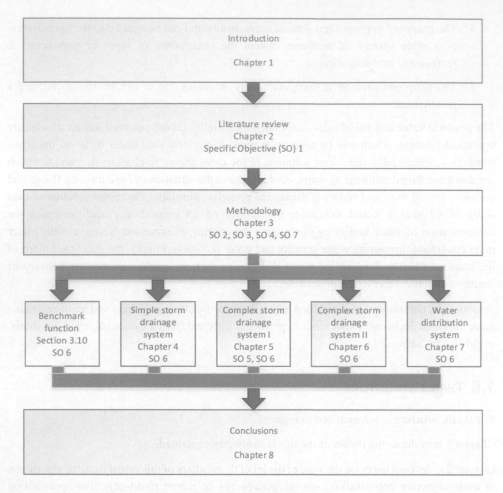

Figure 1. Thesis structure

In Chapter 4 the methodology is applied to the robust optimisation of a simple storm drainage network. This network is optimised by using two approaches, namely deterministic and robust (i.e. ROPAR). The main objective here is to compare the robustness of a deterministic solution with the robustness of a ROPAR solution.

In Chapter 5 the methodology is applied to two complex storm drainage networks. Both networks are optimised using two robust approaches, namely OSOF and ROPAR. The main objective is to compare the robust solutions generated by both approaches.

In Chapter 6 the methodology is applied on a more complex storm drainage system. This system is optimised using two approaches, deterministic and robust (i.e. ROPAR). The main objective is to test ROPAR with a problem with more than two objective functions and more than one uncertainty source.

Chapter 7 shows the application of the methodology to four water distribution systems of different topology and size. These systems are optimised deterministically by a new optimisation algorithm, and robustly by using ROPAR. The main objective is to test ROPAR with this new deterministic optimisation algorithm and with 24 uncertainty sources representing the hourly water demand.

Finally, Chapter 8 presents the conclusions, where answers to the research questions are shown, as well as the outlook and recommendations to continue this research further.

2

LITERATURE REVIEW

This chapter presents review of literature on multi-objective optimisation algorithms and multi-objective robust optimisation. This review leads to identification of unresolved issues regarding the latter as well as formulation of open research questions.

2.1 EVOLUTIONARY ALGORITHMS

Traditional optimisation methods used in operations research such as linear programming and nonlinear programming are capable of finding optimal solutions for a wide variety of problems which objective function(s) and constraints can be expressed analytically. However there are many problems for which analytical representation is not possible since are expressed in the form of a computer programme (code). For such problems it is possible to calculate the value of an objective function for a given vector of decision variables, but not directly the gradient of the objective function, so the efficient gradient-based methods cannot be used. For such problems the methods of direct search and heuristic optimisation, e.g. evolutionary algorithms (EA), are used.

EAs are loosely based on the theory of evolution by Darwin (1859). From this theory, four principles are drawn to simulate biological evolution of the candidate solutions (Chiong et al. 2012):

1. A set of individuals (i.e. candidate solutions) form a population.

2. The population is constantly changing because new individuals (i.e. births) are included in the population while others are discarded (i.e. deaths).

3. A measure of fitness is used to decide which new individuals are included in the population and which other individuals are discarded.

4. There is a mechanism to 'reproduce' individuals in order to generate new individuals. These new individuals are similar to their parents although not exactly the same.

These principles are the basis of the typical cycle appearing in every EA. This cycle is shown in Figure 2; in it, the four principles are repeated until a stopping criterion is met. Next, the main parts of the cycle are explained.

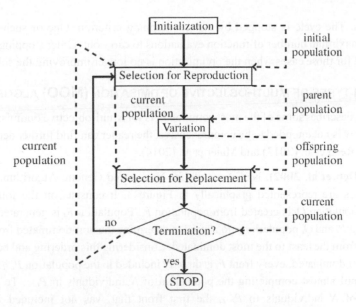

Figure 2. Typical cycle of an Evolutionary Algorithm. Solid arrows show control flow, dashed arrows show data flow (Jansen 2013)

Initialization. EAs begin with an initial set of solutions, which typically are random. This set of solutions is named initial population. Each solution, or more properly said, candidate solution of this population is also named an individual. Furthermore, for the sake of the next step, this initial population becomes the current population.

Selection for reproduction. Some of the (best) individuals of the current population are selected to be reproduced. These selected individuals form the parent population.

Variation. Using a mechanism of inter-combination, the individuals of the parent population are reproduced in order to generate a new population. This new population is named offspring population.

Selection for replacement. Each new individual is evaluated using the objective functions. For single-objective optimisation, a certain proportion of the best individuals are selected. In multi-objective optimisation, the individuals that are better than the rest of the individuals, with respect to at least one of the objective functions, form a Pareto front. This Pareto front is the result of having trade-off among the objective functions. Some of those individuals not belonging to the Pareto front are eliminated from the current population. Additionally, if these non-dominated solutions are too many, or maybe because they are very similar, then some of these solutions can be eliminated from the current population using a criterion. One of these criteria is measuring the crowding distance of each solution. The crowding distance measures how close are the solutions in the objective space.

17

Termination. The cycle is stopped by using a stopping criterion. One of such criteria is by reaching a maximum number of function evaluations to carry out. Other stopping criterion can also be used for those cases when the optimisation is no longer improving the solutions.

2.2 MAIN TYPES OF MULTI-OBJECTIVE OPTIMISATION (MOO) ALGORITHMS

This section describes some of the most common EAs for Multi-objective optimisation (MOO). As this review is not intended to be comprehensive, the reader can find further details and extra literature in Reed et al. (2013) and Maier et al. (2014).

NSGA II (Deb et al. 2002). NSGA stands for Non-sorted Genetic Algorithm, whose main characteristics are represented graphically in Figure 3. It consists on the following steps. Offspring population Q_t is created from population P_t. Population R_t is generated as the result of combining P_t and Q_t populations. R_t is used to determine the non-dominated fronts F_i. Fronts are ordered from the least to the most dominated. Considering this ordering and beginning with the front least dominated, every front F_i is directly included in the population P_{t+1}. This process continues until almost completing the population of N individuals in P_{t+1}. To complete the population of N individuals in P_{t+1}, the first front that was not included is taken into consideration; those individuals of this front to be included in P_{t+1} are selected using the algorithm to calculate the crowding distance.

Figure 4 shows the graphical representation of crowding distances of stacking. The algorithm calculates the cuboids surrounding each of the solutions. Solutions that have the largest cuboids are selected to achieve greater diversity of solutions included in the population P_{t+1}.

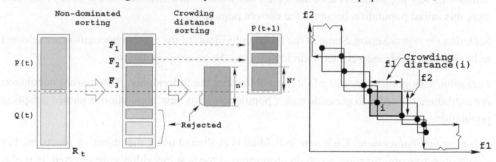

Figure 3. Representation of NSGA II (Deb et al. 2002)   *Figure 4. Crowding distance (Deb et al. 2002)*

Epsilon-MOEA (Deb et al. 2005). This algorithm is a modification of NSGA-II. MOEA stands for Multiple Objective Evolutionary Algorithm. The epsilon concept is one of the basic features of this algorithm; the search space is divided into cells (or hyper-boxes) and diversity is maintained, ensuring that each cell or hyper-box can only be occupied by only one solution. Epsilon represents the interval which is not significant for the user in terms of making a decision. For example, if the cost of a project is around one billion dollars, epsilon could be one hundred thousand dollars. Epsilon is compared with the difference of two values of the same objective function, in the previous example the difference of the cost of two projects.

Another new feature, is that this is a steady-state algorithm. This means that instead of generating the number of individuals equal to the size of the population in each generation, it generates only two individuals. Epsilon-MOEA keeps evolving two populations: the population P_t and the individuals in the file E_t, corresponding to "EA" and "Archive" in Figure 5 respectively.

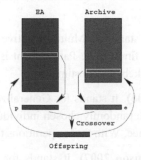

Figure 5. Crossover (Deb et al. 2005)

C-NSGA-II (Deb et al. 2005). C-NSGA-II stands for Clustered Non-dominated Sorting Genetic Algorithm-II. This algorithm is very close to NSGA II. The main difference lies in the way it makes the selection of individuals from the first front that was not included. Instead of using crowding distance, it uses a clustering algorithm.

rNSGAII (Kapelan et al. 2006). rNSGAII stands for robust NSGAII. rNSGAII is a modification of NSGAII to find robust optimal solutions. The idea is that the robustness of each chromosome is tested in each generation. The robustness of a chromosome in the current generation is assessed also considering its robustness in previous generations.

Epsilon NSGA II (Kollat and Reed 2006). Epsilon-NSGAII adds features to the NSGAII algorithm: epsilon-dominance archiving, adaptive population sizing, and automatic termination. Epsilon-dominance is the same concept of epsilon in epsilon-MOEA described earlier.

With respect to adaptive population sizing, the population size in each generation is adapted according to the complexity of the problem. For each generation, a quarter of the individuals will be taken from the epsilon-non-dominated solutions and the remaining three quarters of the individuals will be randomly generated.

The automatic termination may come from two user-specified criteria. The first criterion is due to conditions such as reaching a maximum run duration or not meeting a rate of diversity. The second criterion is used when the number of solutions is not growing according to the specified rate of change.

OMOPSO (Reyes-Sierra and Coello-Coello 2005). OMOPSO stands for Optimal Multi-objective Optimisation PSO. It is a modification to the Particle Swarm Optimisation algorithm (PSO) to handle multiple objectives. Each non-dominated solution is considered as a new leader, using a crowding factor to control the size of the population. Also, it uses the concept of epsilon dominance to control the number of reported solutions.

IBEA (Zitzler and Kunzli 2004). Its name stands for Indicator-Based Evolutionary Algorithm and as the name suggests, it is based on indicators. Each indicator is generated from each objective function via a transformation to normalize its range. All the indicators are used as arguments of the fitness function.

MOEA/D (Zhang et al. 2009). It stands for Multi-Objective Evolutionary Algorithm based on Decomposition. The problem of finding the Pareto front is divided into a number of scalar optimisation problems using a Tchebycheff estimator.

GDE3 (Kukkonen and Deb 2006). It stands for Generalized Differential Evolution. It uses the scaled difference between two randomly chosen individuals. A third random individual is chosen to be added to the difference. With these operations the new individual is generated.

AMALGAM (Vrugt and Robinson 2007). It stands for A Multi ALgorithm Genetically Adaptive Method. The cycle of this hybrid algorithm begins with a population k (see Figure 6). The offspring from this population is generated concurrently using four algorithms: Nondominated Sorting Genetic Algorithm II (NSGA-II), Particle Swarm Optimisation (PSO), Adaptive Metropolis Search (AMS), and Differential Evolution (DE). Each offspring i is evaluated to determine which one has a better performance to find the Pareto front. According to this evaluation, every algorithm is weighted to let them generate offspring in the next cycle $k + 1$ in line to its success in cycle k. From the union of the offspring k and the population k the individuals are selected to be part of the new population $k +1$.

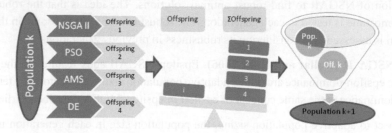

Figure 6. AMALGAM algorithm

Borg (Hadka and Reed 2013). Sharing the same idea used by Amalgam, Borg is also designed using a combination of optimisation methods to improve the search of the optimal Pareto front. Borg assimilates several design principles from existing MOEAs and introduces several novel components. These components include:

- an epsilon-box dominance archive for maintaining convergence and diversity throughout the search;
- epsilon-progress, which is a measure of search progression and stagnation;
- an adaptive population sizing operator based on epsilon-NSGA-II's use of time continuation to maintain search diversity and to facilitate escape from local optima;
- multiple recombination operators to enhance search in a wide assortment of problem domains; and,
- steady-state elitist model of epsilon-MOEA.

AMGA2 (Tiwari et al. 2011). It stands for Archive-based Micro Genetic Algorithm. It is considered a steady-state genetic algorithm because its main Pareto front has a small number of solutions, although other good solutions are kept stored in an archive. To produce the next generation of populations, it uses all the solutions in the main Pareto front mated with some of the solutions in the archive. To decide which solutions to include in the new Pareto front, two criteria are used: first, the degree of dominance of the solution; second, the diversity of the solution. In this way two goals are reached, namely a small number of function evaluations and high diversity. The good solutions that are not selected for the new Pareto front are included in the archive. To maintain the archive, the solutions crowding a specific region of the solution space are eliminated using the nearest neighbour search strategy.

This algorithm addresses some of the weaknesses of other MOEA algorithms. For example, for each new individual, the constraints are evaluated before making an unnecessary evaluation of the objective function. The most important features of AMGA2 are the following:

- it is built in a modular way, enabling the independent improvement of each module;
- it uses a dynamic and small population;
- it uses an external file to store the best solutions;
- the wanted number of solutions is defined before starting the optimisation algorithm; and,
- it uses adaptable dynamic selection of the crossover operator.

2.3 Robust multi-objective optimisation (RMOO)

This section begins stating the definition of robust optimisation of multiple objectives that is guiding this research. This definition is drawn from several definitions in the literature. Then, those works carrying out robust multi-objective optimisation in water sciences are reviewed. The last part is describing those techniques of robust multi-objective optimisation that can be applied to several types of problem without having to modify the algorithmic definition of robustness.

2.3.1 Definitions of RMOO in literature

Explicit consideration of uncertainty in optimisation of multiple objectives is relatively recent research. Table 1 shows definitions of robust optimisation proposed by several authors. They are listed in chronological order since 1990. It can be seen that they are mainly in line with previous discussions.

Table 1. Points of view related to uncertainty in optimisation

Author	Context or definition of robustness
Klein et al. (1990)	"Uncertainty presents unique difficulties in constrained optimisation problems. These difficulties are exacerbated when there exist multiple, often conflicting, criteria. In particular, contemporary multiple criteria interactive methods may not be appropriate under uncertainty because the concept of an efficient frontier is lost."
Ben-Tal and Nemirovski (2002)	"Robust Optimisation is a modeling methodology, combined with computational tools, to process optimisation problems in which the data are uncertain and is only known to belong to some uncertainty set."
Jin and Sendhoff (2003)	"Robustness of an optimal solution can usually be discussed from the following two perspectives: - The optimal solution is insensitive to small variations of the design variables. - The optimal solution is insensitive to small variations of environmental parameters."
Gunawan and Azarm (2005)	"In multi-objective design optimisation, it is quite desirable to obtain solutions that are 'multiobjectively' optimum and insensitive to uncontrollable (noisy) parameter variations. We call such solutions robust Pareto solutions."
Savic (2005)	"Robustness is defined as the ability of the system to maintain a level of performance even if the actual parameter values are different from the assumed values."
Deb and Gupta (2006)	"In practice, users may not always be interested in finding the so-called global best solutions, particularly when these solutions are quite sensitive to the variable perturbations which cannot be avoided in practice. In such cases, practitioners are interested in finding the robust solutions which are less sensitive to small perturbations in variables."
Ong and Lum (2006)	"The global optima may not always be the most desirable solution in many real world engineering design problems. In practice, if the global optimal solution is very sensitive to uncertainties, for example, small changes in design variables or operating conditions, then it may not be appropriate to use this highly sensitive solution."

Author	Context or definition of robustness
Beyer and Sendhoff (2007)	"Robust design optimisation – the search for designs and solutions which are immune with respect to production tolerances, parameter drifts during operation time, model sensitivities and others."
Gaspar-Cunha and Covas (2008)	"Practical solutions to engineering optimisation problems should also be robust, i.e., the performance of the optimal solution should be little affected by small changes of the design variables, or of environmental parameters."
Witteveen and Iaccarino (2010)	"Robust design optimisation has become essential to make high–performance aerospace designs insensitive to uncertainties in the environment and the design parameter."
Petrone et al. (2011a)	"Robust Optimisation is an extension of conventional optimisation procedures and aims at taking in account uncertainty in the design procedure."
Congedo et al. (2011)	"Aim of robust design optimisation is to determine a design which is relatively insensitive with respect to physical and modeling uncertainties."
Petrone et al. (2011b)	"Optimised configurations can become ineffective in the presence of unexpected operating scenarios, for example in the presence of insect or dust contamination, thus requiring recalibration and adjustments. Robust design procedures aim at limiting the potential impact of uncertainties on performance and are an effective risk-mitigation strategy."
Erfani and Utyuzhnikov (2012)	"In design and optimisation problems, a solution is called robust if it is stable enough with respect to perturbation of model input parameters."
Marchi et al. (2014)	"(…), in most cases, optimal solutions found with deterministic approaches are not robust against input parameter variations or have a certain probability to violate constraints or limit state functions."

Author	Context or definition of robustness
Cheng et al. (2015)	"Robust design optimisation has been proposed which aims at obtaining optimal solutions while maintaining the insensitivity of those solutions to uncertainties."
Maier et al. (2014)	"(…) robustness (or the ability to cope if future trajectories of the system are unfavourable)"

In the definition of robust optimisation given in Chapter 3, the definitions cited above have been taken into account.

2.3.2 RMOO in water related systems

The water sector started to use mathematical optimisation algorithms in the 1960s (Karmeli et al. 1968; Schaake and Lai 1969). Nicklow et al. (2009), Reed et al. (2013) and Maier et al. (2014) devoted papers to review this topic, particularly focusing on the use of genetic algorithms. However, although for many years in typical problem settings the issue of uncertainty was not considered in multi-objective optimisation of water systems, in the last 10-15 years this aspect has started to be given attention. This type of optimisation is often (albeit not always) referred to as robust optimisation. The notion of robustness is treated in different studies in various ways and is determined by the particular needs of a case study or the authors' preferences.

Some of the first works applying RMOO methods were in the field of water distribution systems. In these works (Babayan et al. 2004; Kapelan et al. 2005; Savic 2005; Kapelan et al. 2006; Savic 2006; Odan et al. 2015), the robustness of the identified solutions was achieved by including a measure of robustness as one of the objective functions, which depends on the type of problem being solved. One example of such a specific objective was used by Kapelan et al. (2006): measuring the probability that at the same time the heads of all the nodes in the network comply with at least the minimum head required for each node. In contrast, in the approach proposed in this thesis, the robustness is reached not by requiring an objective function measuring the robustness depending on the problem being solved, but by using criteria of general applicability.

Other research in the water area has dealt with the optimum design of urban drainage networks and wastewater systems. All these works use OSOF as the method to accommodate uncertainty. OSOF is explained in more detail in the next section. Andino-Santizo (2012) used two methods to find the robust optimum design of a drainage network considering uncertainty in the population growth. Kang and Lansey (2012), Vojinovic et al. (2014), Yazdi et al. (2014) and Kebede (2014) used a robust optimum design for water, storm- and wastewater infrastructures considering as uncertain the following parameters, respectively: amount of water; climate change, urbanization, population growth, and pipe aging; rainfall; and, Manning roughness.

Galindo-Calderon et al. (2015) found the robust optimum type and location of the Best Management Practices for a storm drainage network.

Yet another approach of RMOO used for water-related systems follows the notion of deep uncertainty. Walker et al. (2013) classified the uncertainty in two categories: first, uncertainties that can be statistically analysed; and second, uncertainties that cannot be statistically analysed due to the unforeseen future (named deep uncertainty). Beh et al. (2015) solved the problem of the optimal sequencing of additional water supply sources over a 40-year planning horizon. Mortazavi-Naeini et al. (2015) optimised planning and operation of bulk water systems foreseeing conditions of extreme drought. Watson and Kasprzyk (2017) optimised water allocation from multiple market-based supply instruments.

It is worth mentioning that, differently from deep uncertainty, in this research uncertainty is assumed to be such that it allows for probabilistic description and analysis.

It is worth mentioning the two relatively recent papers focused on deep uncertainty, which however, contain some parts that could be used to develop further the framework proposed in this thesis. The first paper is by Herman et al. (2015). In this paper several frameworks for robust decision making are analysed. From this analysis a taxonomy is derived by distinguishing four common stages: I) Alternatives, II) States of the world, III) Robustness measures and IV) Robustness controls. The framework in this thesis already has the three first stages. The fourth stage, related to sensitivity analysis, presents ideas that could be in principle also used to extent the framework of this thesis further. The second paper is by McPhail et al. (2018). In this paper a framework is proposed to analyse the common stages to calculate a robustness metric. Then using this framework, eleven robustness metrics are analysed to determine how they are calculated and when it is useful to use each of them. These metrics are also ordered from the highest risk averse to the lowest risk averse. The framework in this thesis uses four out of these eleven robustness metrics, including the first one which is the highest risk averse. The remaining seven robustness metrics could be also considered in the future development of the framework proposed in this thesis.

In the reviewed literature the problem formulations used do indeed take uncertainty into account, but it can be seen that uncertainty is in a way 'hidden' and cannot be directly estimated. The problem of robustness is formulated by adding a specific metric of robustness depending on the considered problem type, and therefore varies in different case studies; besides, the process of probabilistic uncertainty propagation to the final solutions is typically not explicit either. This prompted the development of a new algorithm for robust optimisation with a general applicability. The novelty of this study also relates to addressing features not fully considered by the approaches previously mentioned. First, it is a method where the robustness is not defined by a metric of robustness depending on the type of problem being optimised. Second, this method does not change an objective function, e.g. by its smoothing, employed earlier. Third, the method characterizes the uncertainty explicitly, using probability density functions.

2.3.3 Techniques for RMOO of general applicability

In this thesis, the robustness of the solution is not defined by a particular measure of robustness depending on the problem being solved. On the contrary, the aim is to define a technique that can be used in several kind of problems without redefining each time what is understood as robustness (however allowing of course for taking into account the problem specifics). For this reason, only those techniques of general applicability in the literature are reviewed next. Those techniques can be classified as follows.

- Using the mean of the objective function.
- Using the mean and variance of the objective function.
- Using an additional objective function related to robustness.
- Using additional constraints related to robustness (which indeed may be related to the particular problem specifics).
- Using CDF comparison.

Descriptions of these techniques are explained as follows.

Using the mean of the objective function

Here a simple idea is employed: instead of searching for the minimum of the objective functions f, optimisation is aimed to minimise some sort of a 'robust approximation' of f. One of the possible ways of building such approximation is using a smoothed version of f, e.g. using averages of the objective function values in a proximity of a given point. To obtain these averages, for each individual (i.e. vector, or point) in the population, a set (sample) of points in the vicinity is considered (i.e. set U), for each individual in U the objective function values are calculated, and then the average is found. So instead of minimising f, f_μ is the function aimed to minimise. Mathematically this can be represented as follows:

$$\min_x \ f_\mu\left(x, u\right) \tag{1}$$

$$f_\mu(x, u) = \frac{1}{|U|} \sum_{u_i \in U} f(x, u_i) \tag{2}$$

where f is a vector of objective functions, f_μ is a vector with the 'robust approximation' of f, U is a set containing the points in the vicinity of x, x is a vector of decision variables, u is a vector of parameters with uncertainty, u_i is the i_{th} neighbour.

The motivation behind the technique using the mean of the objective function is that the uncertainty on some parameters causes noisy evaluations of the objective function. To have a consensus on the "right" value of the objective function, all of the noisy evaluations of the objective function belonging to the same "neighbourhood" are averaged. Every author using this technique defines the meaning of the neighbourhood. This approach of calculating f_μ is what is mentioned as the method OSOF.

The following authors used this approach: Andino-Santizo (2012), Deb and Gupta (2006), Fieldsend and Everson (2005), Gaspar-Cunha and Covas (2008), Jin and Sendhoff (2003), Kang and Lansey (2012), Kapelan et al. (2005), Kebede (2014), Kuzmin (2009), Savic (2005), Vojinovic et al. (2014), and Zeferino et al. (2012).

Using the mean and variance of the objective function

This approach is similar to the previous one, but besides calculating the objective function averages f_μ (OSOF) in the proximity of the point in question (set U), the standard deviation σ of the values of f in U is also estimated. The mathematical formulation of the optimisation is as follows:

$$\min_{x} f_{\mu\sigma}(x, u) \tag{3}$$

$$f_{\mu\sigma}(x, u) = A\, f_\mu(x, u) + B\, f_\sigma(x, u) \tag{4}$$

$$f_\sigma(x, u) = \sqrt{\frac{1}{|U|} \sum_{u_i \in U} \left(f(x, u_i) - f_\mu(x, u)\right)^2} \tag{5}$$

$$A + B = I$$
$$a_{jj} \geq 0, b_{jj} \geq 0, \forall j$$

where f is a vector of objective functions, f_μ is the same formulation as Equation (2), f_σ is the 'σ-robust approximation' of f using standard deviations of the objective functions, A and B are diagonal matrices of weights, I is the identity matrix, a and b are elements of the matrices A and B, respectively, x is a vector of decision variables, u is a vector of parameters with uncertainty.

The following authors used variants of this approach: Fieldsend and Everson (2005), Gaspar-Cunha and Covas (2008), Jin and Sendhoff (2003), Kang and Lansey (2012), and Zeferino et al. (2012).

Using an additional objective function related to robustness

In this approach, a vector of objective functions (f_r) which represents some measure of the solution robustness is simply added to the original vector of objective functions f. This measure of robustness f_r could be average f_μ (OSOF) or standard deviation f_σ or their combination $f_{\mu\sigma}$

or another way of measuring the robustness of the system being optimised. The formulation of the optimisation problem in general case is the following:

$$\min_{x}[f(x,u), f_r(x,u)] \tag{6}$$

where f is the vector of the original objective functions, f_r is an additional vector of objective functions measuring the robustness of the solution, x is the vector of decision variables, and u is a vector of parameters with uncertainty.

This approach was used by Babayan et al. (2004), Erfani and Utyuzhnikov (2012), Kapelan et al. (2006), Marchi et al. (2016), Roach et al. (2016), Toloh (2014), and Zeferino et al. (2012).

Using additional constraints related to robustness

In a yet another approach, a vector of constraints is added to the original inequality constraints of the problem, which sets a boundary B to a given measure of robustness. Just as in the previous subsection, the robustness could be expressed by any measure f_r defining it. The added constraints would be expressed as

$$f_r(x,u) \leq B \tag{7}$$

where f_r is a vector of functions measuring the robustness of the solution, B is a vector with the value of the robustness that the solution x must have, x is a vector of decision variables, u is a vector of parameters with uncertainty.

The authors using this technique are Deb and Gupta (2006) and Gunawan and Azarm (2005).

Comparing the cumulative distribution functions

As in the previous cases, the objective function f depends on a deterministic variable x as well as on a random variable u; therefore f is not deterministic anymore and the corresponding Cumulative Distribution Function (CDF) $F(x,u)$ can be defined. Petrone et al. (2011a) introduce the **Robustness Index** defined as

$$Robustness\ Index(x) = \int_{U} |F(x,u) - \tau(x)|\ du \tag{8}$$

where $\tau(x)$ can be thought of as the expected response of the system, defined for instance by inverse design or by an ideal design. **Robustness Index** indicates how close is the CDF $F(x,u)$ of an evaluated solution to the ideal CDF $\tau(x)$ of an intended solution. The robust optimisation problem can be then defined as

$$\min_{x} Robustness\ Index(x) \tag{9}$$

2.4 CONCLUSIONS

It can be seen that all the mentioned approaches first aggregate uncertainty into the objective functions or constraints, and then solve a standard MOO problem resulting in a single Pareto front. This is indeed a valid approach to deal with uncertainty, however, it should be noted that the impact of uncertainty is embedded in this Pareto front, so uncertainty is in a way hidden, 'encoded' in the solutions which are expected to be robust. The uncertainty is encoded because the resulting Pareto front combines the minimisation of the objective functions and the robustness in one single piece of information. The problem is that it is not straightforward to 'decode' the level of robustness from the identified solutions and to analyse how Pareto-optimal sets depend on actual realizations of uncertain variables. Decoding this information requires information to be stored for each solution in the Pareto front regarding the relation between the solution and the uncertain parameters. If explicit decoding and presentation of uncertainty are not performed explicitly, this makes it difficult to fully grasp the impact of the uncertainty by the decision makers and to select one of the solutions from the Pareto front.

The extended literature review makes it possible to confirm that existing methods do not necessarily allow the *explicit* propagation of uncertainty from inputs or parameters to solutions, so that development of a new approach would be beneficial.

The following is a summary of the points just mentioned:

1. Most of the current approaches to robust optimisation of multiple objectives produce a single Pareto front, or just several sets based on a limited set of scenarios.

2. The uncertainty of the solutions in this Pareto front is not explicitly represented.

3. Analysis of propagation of the probabilistically expressed parameter uncertainties to the solutions is not explicitly performed.

4. Robust optimisation is typically carried out with respect to one definition (criterion) of robustness (or reliability), and not posed as a multi-objective robust optimisation problem, where various aspects of robustness would be considered as (competing) criteria.

3

METHODOLOGY[1]

This chapter defines the methodological framework, or robust optimisation strategy, classifies possible sources of uncertainty and states the definition of a robust optimisation problem. In addition, it introduces and explains the ROPAR algorithm (Robust Optimisation and Probabilistic Analysis of Robustness) in detail, and defines the experimental setup to test ROPAR.

---

[1] The results of this chapter were published in the following papers:

Marquez-Calvo O. O. & Solomatine D. P. 2019 Approach to robust multi-objective optimization and probabilistic analysis: the ROPAR algorithm. Journal of Hydroinformatics 21, 427-440 doi:10.2166/hydro.2019.095.

Marquez-Calvo O. O., Quintiliani C., Alfonso L., Di Cristo C., Leopardi A., Solomatine D. P. & de Marinis G. 2018 Robust optimization of valve management to improve water quality in WDNs under demand uncertainty. Urban Water Journal 15, 943–952 doi:10.1080/1573062X.2019.1595673.

3.1 METHODOLOGICAL FRAMEWORK

In summary, the steps of the proposed framework are the following:

1. Study environment.

2. Define the optimisation problem.

3. Identify and characterize uncertainty sources.

4. Select the (deterministic) multi-objective optimisation algorithm.

5. Apply the robust optimisation algorithm (ROPAR).

6. Determine reliability of the solutions, if there are constraints on it.

7. Analyse the results, draw conclusions.

Figure 7 shows more explicitly the respective workflow. ROPAR is broken down into several steps: sampling of uncertain parameters (inputs); identification of a Pareto front for each sample by multiple runs of a MOO algorithm; analysis of Pareto fronts; finding the robust solution.

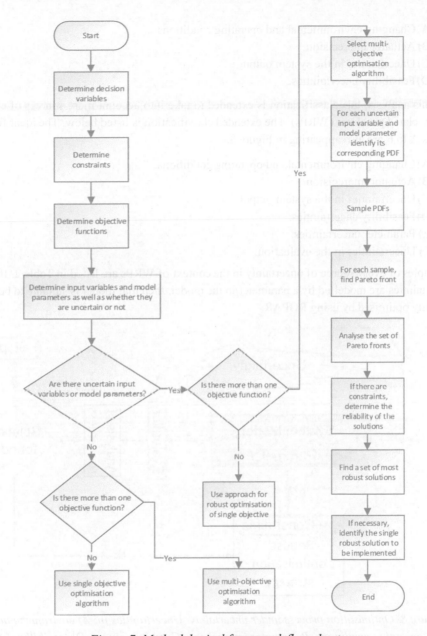

Figure 7. Methodological framework flow chart

3.2 IDENTIFICATION AND CHARACTERIZATION OF UNCERTAINTY SOURCES

According to Beyer and Sendhoff (2007), the sources of uncertainty have the following classification.

33

(A) Changing environmental and operating conditions.

(B) Actuator imprecision.

(C) Uncertainties in the system output.

(D) Feasibility uncertainties.

For this analysis, this classification is extended to take into account more sources of error in water related problems (WRPs). The extended classification is listed below. The identification letters A to F are also appearing in Figure 8.

(A) Changing environmental and operating conditions.

(B) Actuator imprecision.

(C) Uncertainties in the system output.

(D) Feasibility uncertainties.

(E) Parameter uncertainties.

(F) Uncertainties in the evaluation.

Examples of every source of uncertainty in the context of WRPs are listed in Table 2. If such uncertainties are modelled by a parameter in the model, then that model is suitable to be robustly optimised by using ROPAR.

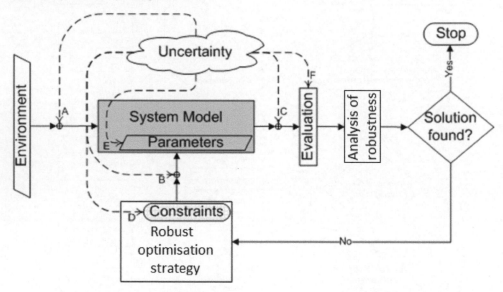

Figure 8. Optimisation process under uncertainty. Uncertainties in: A) environmental and operating conditions, B) actuator imprecision, C) system output, D) feasibility, E) parameters, F) evaluation

Table 2. Examples of uncertainties in water related problems (WRPs)

Type of uncertainty	Examples of uncertainties in WRPs
(A) Changing environmental and operating conditions	Amount of precipitation. Failures in the hydraulic infrastructure.
(B) Actuator imprecision	Pump not delivering the nominal discharge. Automated valve not calibrated or not accurate enough.
(C) Uncertainties in the system output	Measuring devices not precise enough or not reliable.
(D) Feasibility uncertainties	Shortfall of the minimum amount of electrical energy to be generated by a hydropower station.
(E) Parameter uncertainties	Initial conditions of a model, or parameters of the model.
(F) Uncertainties in the evaluation	Inaccuracy of the model to predict water scarcity.

Once the uncertainty source has been identified, the next step is to characterize it. In this step the PDF representing the variability of the uncertainty source is determined. The nature of the PDF is not restricted in any sense.

If the uncertainties cannot be statistically analysed due to the unforeseen future or because insufficient quantity and quality of data available, then a different approach to robust optimisation has to be followed. This kind of uncertainty is usually named 'deep uncertainty'. Some examples of robust optimisation with deep uncertainty are Beh et al. (2015), Mortazavi-Naeini et al. (2015) and Watson and Kasprzyk (2017). Frameworks for robust optimisation with deep uncertainty can be found in Herman et al. (2015) and McPhail et al. (2018). Again, in this thesis we assume the uncertainty can be expressed probabilistically.

3.3 GENERAL FORMULATION OF THE OPTIMISATION PROBLEM

Based on literature review and considerations above, the following definition for Robust Optimisation is adopted:

> *Robust optimisation of multiple objectives is an optimisation method modified in such a way that it generates a solution or a set of solutions that have minimum (or limited) variability of the objective functions when some elements or parameters of the modelled system vary due to their uncertainty.*

35

This definition is the cornerstone of all of the research developed in this work. This verbal definition is complemented with the following mathematical definition of a problem of robust optimisation of multiple objectives, and with the mathematical definition of *robustness metrics* in Section 3.4.

The definition of robust optimisation above stated can be formalized mathematically for the case of a RMOO as follows. Consider a system model, which is an abstraction (either mathematical or computational) of a real system intended to be optimised. The behaviour of the system model is characterized by the value taken by the input uncertain variables u, and decision variables x. A solution is a configuration or a design that is defined by a specific combination of the values of the decision variables x.

With this in mind, a robust multi-objective optimisation problem can be formulated as

$$\min_{x} f(x, u) = \min_{x} f_i(x, u) \; i = 1, 2, \dots, n \tag{10}$$

subject to:

$$g_j(x, u) \leq 0 \; j = 1, 2, \dots, k \tag{11}$$

$$h_j(x, u) = 0 \; j = 1, 2, \dots, q \tag{12}$$

$$x^{min} \leq x \leq x^{max} \tag{13}$$

$$u^{min} \leq u \leq u^{max} \tag{14}$$

where f is the vector of the n objective functions ($f_1, f_2, \dots, f_n$), g_j is the j-th inequality constraint, h_j is the j-th equality constraint, k is the number of inequality constraints, q is the number of equality constraints, u is the vector of the uncertain input variables (e.g. rainfall), u^{min} and u^{max} are the vectors of the minimum and maximum values of uncertain input variables, x is the vector of the decision variables (e.g. pipe diameter), x^{min} and x^{max} are the vectors of the minimum and maximum values of the decision variables, and $f_i, g_j, h_j, x, u \in \mathbb{R}$.

Solutions of a multi-objective optimisation problem are typically represented as a Pareto front F in the space of objectives (criteria) which can be defined as

$$F \equiv \{f_q\}, q = 1, 2, \dots, ns \tag{15}$$

where f_q is the q-th tuple representing the evaluations of the n objective functions in the Pareto front, F are all the evaluations of the objectives functions comprehending the Pareto front, and ns is the number of tuples comprising the Pareto front.

To summarize, a solution is a configuration or a design that is defined by a specific combination of the values of the decision variables x. The aim is to find a robust optimum solution x which makes the objective functions f remain near an optimal value regardless of the uncertainty that could affect the parameters u.

3.4 MATHEMATICAL DEFINITIONS OF ROBUSTNESS METRICS

In this research, the robustness criteria are based on the general definitions of robustness. These four criteria are based on the four most common robustness metrics used in the literature (Beyer and Sendhoff 2007). These four metrics of robustness are exemplified below. Assume that the goal is to minimise the objective function $f(x)$. Consider that its robust counterpart is the function $F(x)$.

The first metric of robustness is the Expected Value (denoted with subindex 'ev') of f, which can be represented by:

$$F_{ev}(x) = \int f(x, u)\, p(u)\, \mathrm{d}u \qquad (16)$$

where $p(u)$ is the probability density function of the uncertain variable u.

The second metric of robustness is generally known as the 'Worst Case' (denoted with subindex 'wc'). This is represented by:

$$F_{wc}(x) = \sup_{V \in X(x,u)} f(V) \qquad (17)$$

where V is the neighbourhood of the solution x.

The third metric of robustness is the Standard Deviation (i.e. sd) of f represented by:

$$F_{sd}(x) = \sqrt{\int (f(x, u) - f(x))^2 p(u) \mathrm{d}(u)} \qquad (18)$$

The fourth metric of robustness is also known as 'Probabilistic Threshold' (i.e. pt). Assume the following two conditions: first, there exist the threshold of interest q; second, one is looking for the solution x where the probability of f being less or equal than q is a maximum. This can be expressed by:

$$F_{pt}(x) = \Pr(f \le q|x) \to \max \qquad (19)$$

The fourth metric in fact represents *reliability* of a system, as understood by Loucks and Van Beek (2017) and Jin (2019). Loucks and Van Beek (2017) define it as

'The *reliability* of any time series can be defined as the number of data in a satisfactory state divided by the total number of data in the time series'.

Jin (2019) defines it as

> '*Reliability* is defined as the ability of a system or component to perform its required functions under stated conditions for a specified period of time … It is often measured as a probability of failure or a possibility of availability'.

The latter definition allows to consider the fourth metric as a measure of *reliability* even though it is not expressed as a probability or a ratio.

Reliability also has a close connection to the notion of robustness used by Kapelan et al. (2005), Savic (2005), Kapelan et al. (2006) and Savic (2006), where robustness of water distribution systems is defined as the percentage of nodes with a pressure above a certain threshold. Actually, reliability has been often used interchangeably as robustness in water sciences, because, indeed, it is one of the ways of measuring robustness.

The presented four robustness metrics reflect quite important aspects of robustness in the optimisation context. At the same time, various other approaches to robust optimisation in the literature may use different ways of handling robustness, and may have various ways of formulating its indicators (criteria). It is worth mentioning that in case of requiring other formulations of robustness, they can easily be integrated in the considered ROPAR framework as well.

The robustness metrics used in this thesis are useful for the following reasons. F_{ev} and F_{sd} relate to robustness because we want to find a solution x such that the possible values of $f(x,u)$ are similar, regardless the value of u. F_{ev} and F_{sd} aim at objectives alike, although not the same, because they are using the first and second statistical moment of the random variable $f(x,u)$, respectively. F_{wc} relates to robustness because we want to find a solution x that is going to produce the value of $f(x,u)$ as the least worst regardless the value of u. Finally, F_{pt} relates to robustness because we want to find that solution x that is going to produce the least number of cases where $f(x,u)$ reaches an undesired threshold q regardless the value of u.

These four metrics are implemented in ROPAR algorithm presented in Section 3.6.1.

3.5 SELECTION OF MOO ALGORITHM

3.5.1 MOO algorithm for the storm drainage cases

From the algorithms revised in Section 2.1, NSGAII and AMGA2 were chosen to be used in the experiments of this thesis. NSGAII was used in the smallest case study because despite not being the most recent algorithm, it is still the most used nowadays. AMGA2 was used in the rest of the case studies (except the water distribution case studies) because in performance tests carried out to compare the algorithms, it was one of the most efficient algorithms; besides, it is readily available, well documented and structured, and can be easily modified to serve our needs.

In the choice of an optimisation algorithm, the efficiency is very important. This is especially significant in those cases where the value of an objective function is the value generated by a hydrological model that might take minutes, hours, or even days to run.

3.5.2 MOO algorithm for the water distribution cases

The first intention was to use either AMGA2 or NSGAII to solve the optimisation problem regarding the minimisation of water age in a WDS. However, neither of these two optimisers were adequate nor any other of the algorithms revised in Section 2.1. This is due to the fact that the considered problem has a large number of 'engineering' constraints, and these randomized search optimisers produce a large number of unfeasible solutions, stalling the optimisation process. For this reason, it was necessary to develop a specialized adequate optimiser. Chapter 7 describes the proposed optimiser algorithm and how this optimiser was used along with ROPAR to find a robust solution.

3.6 FORMULATION OF THE PROPOSED RMOO ALGORITHM

3.6.1 The ROPAR algorithm

ROPAR stands for Robust Optimisation and Probabilistic Analysis of Robustness. ROPAR is the approach used to find solutions that comply with the definition of robust optimisation in Section 2.3.1. This approach is based on Monte Carlo sampling of the input uncertain parameter u (see definition of u in Section 3.3). It allows to analyse the robustness of solutions found by MOO algorithms.

ROPAR has four main parts. The first part is sampling of the uncertain parameter(s). For each of the samples, a model of the problem is defined with the sampled parameter. In the second part, each of these models of the problem is deterministically optimised separately, thus leading to a Pareto front. Therefore, at the end, there are as many Pareto fronts as the number of samples. The third part of ROPAR consists of analysing the Pareto fronts. Using a visual approach, the Pareto fronts are examined to determine a 'level' at which one objective function exhibits low variability (i.e. low uncertainty). The fourth part of ROPAR is finding the robust solution. Using analytical approach, the four criteria discussed above (Section 3.4) are used to find robust solutions.

The pseudocode of the ROPAR algorithm, highlighting its main four parts, is presented next.

Pseudocode of the ROPAR algorithm

Part 1 (sampling)

1. for $r = 1$ to np do
   begin
2.   Sample u_r

Part 2 (generation of the Pareto front)

3.   Find the Pareto-optimal set F_r by solving the deterministic multi objective
   optimisation problem for the sampled u_r.
   end.

Part 3 (visual analysis of Pareto fronts, and building empirical distribution for a user-defined level of one OF)

4. Pick one objective function f_k, say f_2. Choose a certain level L_2 for f_2 (represented
   by some narrow interval for f_2) and form the solutions set S with the values of
   f_2 in this interval, taking solutions from all Pareto sets F_r where $r=1,..,np$,.
5. Pick another objective function, say f_1.
6. Build the empirical distribution of the values of f_1 corresponding to the members of
   set S. This empirical distribution can be approximated by a probability density
   function characterising the uncertainty of f_1 for the solutions corresponding to
   the chosen level L_2 of f_2 (see Figure 9).
7 (optional). Repeat steps 4, 5 and 6 running through a number of various levels L_2,
   until finding a level of f_2, for example the one with minimum variance of f_1.

Part 4 (finding the robust solution(s))

8. Once a level L_2 has been chosen, create the set S containing optimum solutions with
   a value of f_2 approximately equal to L_2. S should include one solution from
   every F_r where $f_2 \approx L_2$ (the closest one). Optionally S could include not one but
   several solutions from every F_r where $f_2 \approx L_2$.
9. To select a set of robust solutions, first measure the *RobustnessIndicator* of every
   solution s in S. *RobustnessIndicator* of the solution s is defined by this
   quadruple:

$$RobustnessIndicator(s)$$
$$= \big(Meanf1(s), Maxf1(s), StdDevf1(s), NonZerof1(s)\big) \qquad (20)$$

Criterion 1 ('expected value' approach):

$$Meanf1(s) = \left[\frac{1}{np}\sum_{r=1}^{np} f_1(s, u_r)\right] \tag{21}$$

Criterion 2 ('worst case' approach):

$$Maxf1(s) = f_1(s, u_q) \mid f_1(s, u_q) \geq f_1(s, u_k)\ q, k \in [1..np]\ \forall k \tag{22}$$

Criterion 3 ('standard deviation' approach):

$$StdDevf1(s) = \sqrt{\frac{1}{np}\sum_{r=1}^{np}\left(f_1(s, u_r) - Meanf1(s)\right)^2} \tag{23}$$

Criterion 4 ('problematic events' (or reliability) approach):

$$NonZerof1(s) = |\{f_1(s, u_k)|f_1(s, u_k) > 0\}|\ k \in [1..np] \tag{24}$$

The set of the robust solutions *s* are found by solving a four-objective minimisation problem, given that the *RobustnessIndicator* has four elements (objectives), being the search space all solutions in *S*, or mathematically:

$$\min_{s} RobustnessIndicator(s)\ s \in S \tag{25}$$

If the measures in *RobustnessIndicator* can be aggregated into one, it will allow for identifying the single most robust solution (see next section).

10. (optional). Repeat steps 8 and 9 running through a number of various levels L₂, as many times as the decision maker needs.

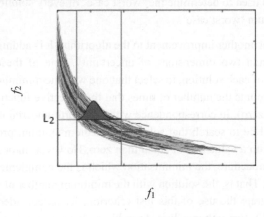

Figure 9. Illustration to the ROPAR Step 6

Please note that up to Step 8, f_1 was calculated for every solution s but only for one of the sampled u_r, so calculation of $f_1(s, u_r)$ for $r = 1,..., np$, in Equations (21)-(24) will actually require calculation of f_1 for np-1 times more, and for each of these this would require running the main model. This results in an additional computational effort.

It is also worth noting the following three points. First, in the optional Step 7, the goal is to scan the robustness of the solutions for several values of f_2. It is just an initial scan because the robustness of the solutions at different values of f_2 is quantitatively defined after carrying out the steps 8, 9, and 10. Second, the minimisation represented by Equation (25) is not a full-fledged optimisation, because the aim is simply to identify that solution s (or those s's) in S that is (are) not dominated. Third, each of the four elements of *RobustnessIndicator* (*RI*) is a different measure of variability, because of this, the most robust solution is the one with the minimum *RI*, in accordance with the definition of Section 2.3.1.

ROPAR can be seen as a family of algorithms, and has different names that represent the evolution of this algorithm. Those names are also useful to identify the particular version of the algorithm used to solve a case study. However, it is worth mentioning that when the word ROPAR without quotation marks is used, it is referring to the algorithm in general, not to a specific version of the algorithm. In this thesis, some of the case studies are solved using 'ROPAR AD2' and some are solved using 'ROPAR AD4'.

'ROPAR A' was first presented by Solomatine (2012). This initial version are the steps 1-7 of the algorithm. The purpose of this version was (and still is) to analyse the propagation of the uncertainty to the solutions of the optimisation.

'ROPAR AD2' is an improved version, adding the three important steps to the original version. These are the steps 8 and 9 and 10, with only two criteria (Criterion 1 and Criterion 2). The purpose of this version is to find the most robust solution. To carry out such task, Criterion 1 determines the 'expected value' of each solution and finds the one with minimum 'expected value'. Criterion 2 is used to determine the 'worst case' of every solution, and then it finds the solution with minimum 'worst case'.

'ROPAR AD4' is yet another improvement to the algorithm. It is adding two criteria to Step 9. These two criteria add two dimensions of uncertainty. One of these criteria measure the 'standard deviation' of each solution, to select that one with the minimum 'standard deviation'. The other criterion counts the number of times that the objective function of relevance is zero (i.e. the threshold is zero). In correspondence with the fourth uncertainty metric (i.e. Equation (19)), the aim would be to search that solution with the maximum probability of having the objective function with a value less or equal than zero. However, in order to have consistency in the optimisation procedure, the minimisation which is the complement of the maximisation is looked for instead. That is, the solution with the minimum number of values bigger than zero is selected. To illustrate the use of this last criterion, let us consider as objective function measuring the flood water volume. It is desirable to have a solution that has the minimum number of scenarios where the flood water volume is different from zero. In this particular context, 'different from zero' is used instead of the more correct term 'bigger than zero' because there is no negative flooding. The term 'different from zero' or simply 'non-zero' is preferred in this thesis because is shorter and more understandable.

Summarizing the four criteria, the first criterion aims to select a solution that minimises the expected value of the objective function (OF); the second criterion – to minimise the largest possible value of the OF (the worst case); the third criterion – to minimise the standard deviation; and the fourth criterion – to minimise the number of problematic (undesirable) events. Generally speaking, all four criteria of step 9 are aiming to find a solution bringing minimum possible variability or change in OF, assuming given uncertainty of parameters or inputs u. This goes in accordance with the definition of RMOO established in Section 2.3.1.

3.6.2 Finding the single most robust solution

As it is pointed out above, the minimisation mentioned in Equation (25) does not require running optimisation algorithms - it is simply to identify that solution s (or those s's) in S that is (are) not dominated. To carry out this minimisation there are two alternatives: first, to find the Pareto front (which would be presented to decision makers for the final selection); or second, aggregate the (normalized) values of the four elements of the quadruple *RobustnessIndicator* to directly select the solution with the minimum aggregated value. Aggregation can be achieved either by linear weighting, or by measuring the Euclidean distance to the origin in the space of four criteria (which is non-linear weighting). Aggregation by simple linear weighting is employed for two reasons: first, it eases the graphical comparison of the solutions; and second, it allows for a straightforward comparison of several optimisation techniques.

The formula of the normalized uncertainty of the solution s is as follows:

$$NormRobustnessIndicator(s)$$
$$= NormMeanf1(s) + NormStdDevf1(s) \qquad (26)$$
$$+ NormMaxf1(s) + NormNonZerof1(s)$$

where $NormMeanf1(s)$ is the result of the division of $Meanf1(s)$ by the maximum value of $Meanf1$ in S. The other three elements of the summation are calculated in a similar fashion. $NormRobustnessIndicator$ is in the interval [0,4]; the lower this value the higher robustness of the solution is.

3.6.3 Determining sample size and confidence level

To estimate the required number of samples, the equation developed by Cochran (1977) is used:

$$n = \frac{z^2 p(1-p)}{e^2} \qquad (27)$$

where n is the number of samples, z is the abscissa of the normal curve that cuts off an area α at the tails ($1 - \alpha$ equals the desired *confidence level*), e is the aimed level of precision, and p is the estimated variability of population, for details see Israel (1992).

3.6.4 Determining the sampling confidence for more than one source of uncertainty

For those cases where there are more than one source of uncertainty u_m (i.e. m_{th} random variable) and considering that they are independent random variables, the *sampling confidence* (i.e. *SC*) of the whole process of sampling u can be calculated. The *SC* is dependent on the *confidence level* of each of the M random variables. The formula relating these concepts is following:

$$SC(\boldsymbol{u}) = \prod_{m=1}^{M} confidence\ level(\boldsymbol{u_m}) \qquad (28)$$

3.7 RELIABILITY OF SATISFYING CONSTRAINTS

One aspect of *reliability* was already used in Criterion 4 (i.e. Equation (24)). In this section yet another aspect of reliability is considered. In Criterion 4 the reliability of the solutions is measured based on the value taken by the objective functions, and in this section the reliability of the solutions is measured based on the value taken by the constraints.

The *reliability of satisfying constraints* (R) of a solution is the ratio between the number of cases where the solution complies with the constraint of the problem (e.g. minimum pressure for every node for a case of water distribution networks) and the total number of cases where the constraint is evaluated. Actually, the engineering concept of *reliability* (Loucks and Van Beek 2017; Jin 2019) is also particularly used in some water distribution problems (Kapelan et al. 2005; Savic 2005; Kapelan et al. 2006; Savic 2006). For the purposes of this section, *reliability* can be represented with the following equation:

$$R = \frac{number\ of\ cases\ where\ the\ contraint\ is\ fulfilled}{total\ number\ of\ cases\ where\ the\ contraint\ is\ evaluated} \qquad (29)$$

The concept of *total reliability* (i.e. *TR*) of a solution can be also introduced. This parameter indicates the minimum performance that we can expect from the solution. So far, two concepts related to probability have been mentioned, *SC* and *R*. These two concepts are used to define the *TR* of a solution. Multiplication of *R* of a solution by *SC* of the sample defines the *TR* of the solution:

$$TR = SC * R \qquad (30)$$

This indicator is used in Chapter 7.

3.8 DEALING WITH MORE THAN TWO OBJECTIVE FUNCTIONS

There are some considerations to take into account for the problems with more than two objective functions. However, before describing such considerations, some nomenclature has to be introduced. Regarding Step 4 of ROPAR, that objective function f_2, which is initially selected, is named *pivotal objective function*. The rest of the objective functions, namely f_1, f_3, $f_4, \ldots, f_n$ where n is the total number of objective functions, are named *nonpivotal objective functions*. The goal of this naming is to make clear that the analysis of robustness is centred on f_2.

Considerations:

1. Steps 5, 6 and 7 of ROPAR are carried out for each possible pair pivotal/nonpivotal functions. At the end of having analysed every pair, one level L_2 of f_1 is selected. Of course if other levels are also of interest, they can be also analysed, however for simplicity, here it is assumed that just one is selected.

2. For every pair pivotal/nonpivotal functions the steps 8, 9 and 10 are carried out. At the end of this analysis, the robustness of all the solutions will be determined and moreover, each solution has as many evaluations of robustness as possible pivotal/nonpivotal pairs exist, let's say P pairs. Each of these evaluations is a different perspective of robustness for a single solution, so that we can say that every solution has P dimensions, or characteristics, of robustness.

3. To select one solution as the most robust, the three options exist.

Options:

3.1. The first option is to select that solution with minimum Euclidean distance to zero in the space with P dimensions. In order to calculate the Euclidean distance, it is suggested to normalize the robustness metrics with the process described in a previous section entitled 'Finding the single most robust solution'.

3.2. The second option is to choose one solution that is the most robust in one of the P dimensions, regardless of the position that that solution has in the rest of the dimensions.

3.3. The third option is to select a solution that has 'good' robustness with respect to two (or more) dimensions although its robustness is not that good for the rest of the dimensions. It is worth noting that there is a possibility that there is only one solution which is the most robust in every of the P dimensions.

3.9 EXPERIMENTAL PLAN

To test the framework developed in this thesis, a set of experiments with cases of different complexity are carried out. In general terms, three types of problems are robustly optimised:

first, benchmark functions; second, storm drainage systems; and third, water distribution systems.

The first kind of problem is the well known benchmark function ZDT1 (Zitzler et al. 2000). Although not related to water sciences per sé, it is a function widely used to test MOO algorithms. This benchmark function was modified to include uncertainty in its definition.

The second kind of problem is design of a storm drainage system. The design is mainly focused to the determination of the pipe diameters, although in a more complex case also solved, the amount and location as well as type of Best Management Practices are determined.

The third kind of problem is the optimisation of operational statuses of the valves to improve water quality in the network. Networks with different sizes and topologies are used.

All these experiments are shown in Table 3.

Table 3. Experiments in this thesis

Problem	ROPAR optimisation	Deterministic optimisation	OSOF optimisation	Number of objective functions	Number of sources of uncertainty	Chapter or section
ZDT1	X	X		2	1	3.10
Storm drainage network with 11 pipes	X	X		2	1	4
Storm drainage network with 59 pipes	X	X	X	2	1	5
Storm drainage network with 100 pipes	X	X	X	2	1	5

Problem	ROPAR optimisation	Deterministic optimisation	OSOF optimisation	Number of objective functions	Number of sources of uncertainty	Chapter or section
Storm drainage network with 100 pipes and BMPs	X	X		3	3	6
Water distribution network with 47 pipes		X		2	0	7.1
Water distribution network with 487 pipes		X		2	0	7.1
Water distribution network with 41 pipes	X	X		2	24	7.2
Water distribution network with 47 pipes	X	X		2	24	7.2
Water distribution network with 366 pipes	X	X		2	24	7.2

Problem	ROPAR optimisation	Deterministic optimisation	OSOF optimisation	Number of objective functions	Number of sources of uncertainty	Chapter or section
Water distribution network with 487 pipes	X	X		2	24	7.2

3.10 EXEMPLIFYING ROPAR

This section presents the application of ROPAR for optimisation of a benchmark function with two objective functions and one source of uncertainty, an illustrative case allowing to demonstrate the workings of ROPAR. The results of ROPAR are compared with the base line case without uncertainty optimised by AMGA2 algorithm. The section ends with the analysis of uncertainty propagation towards optimal solutions.

3.10.1 Problem statement

The widely used function ZDT1 (Zitzler et al. 2000) is employed. This benchmark function is often used to test MOO algorithms. The original formulation of ZDT1 is

$$\min_x f_1(x) = x_1 \tag{31}$$

$$\min_x f_2(x) = \left(1 - \sqrt{\frac{x_1}{g(x_m)}}\right) \tag{32}$$

$$g(x_m) = 1 + \frac{9}{10} * \sum_{i=2}^{11} x_i \tag{33}$$

$$0 \le x_i \le 1, for\ i = 1,2,\dots,11 \tag{34}$$

Its formulation is modified to add randomness, specifically the modified definition of f_2 is changed as follows:

$$\min_x f_2(x) = \left(1 - \sqrt{\frac{x_1}{g(x_m)}} * randomfactor\right) \tag{35}$$

3.10.2 Experimental setup

The objective of this experiment is to compare the robust solutions found by 'ROPAR A' with the deterministic solutions.

Function without uncertainty (deterministic approach)

This problem is represented by the equations (31), (32), (33), and (34). The MOO used is AMGA2 (Tiwari et al. 2011), which is set to run until 10,000 function evaluations of ZDT1 are made.

Function with uncertainty (ROPAR approach)

The problem solved here is represented by equations (31), (35), (33), and (34).

Here *randomfactor* is a random variable with a normal distribution with $\mu = 1$ and $\sigma = 0.05$ (*randomfactor* $\sim N(1, 0.0025)$), which was sampled in the interval [0.8384, 1.1789]. A sample within this interval has 99.2% probability of occurrence. The number of samples for this set of experiments is set to 1,000.

Like the setting for the deterministic optimisation, here AMGA2 was used as the MOO as well. It was configured to run until completing 10,000 function evaluations of ZDT1.

3.10.3 Results and discussion

Optimising the function without uncertainty (deterministic approach)

As a result of optimising the benchmark function ZDT1 without uncertainty (i.e. when *randomfactor* is fixed at 1), the Pareto front in (see Figure 10) is generated.

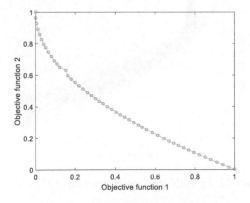

Figure 10. Deterministic optimisation of the ZDT1 benchmark function

Optimising the function with uncertainty (ROPAR approach)

As a result of steps 1, 2 and 3 of the ROPAR algorithm, 1,000 Pareto fronts are generated - see the 1,000 curved lines in Figure 11. In Steps 4-7 of ROPAR, these Pareto fronts are analysed

49

to determine which solutions are most robust within the boundaries imposed by the decision makers.

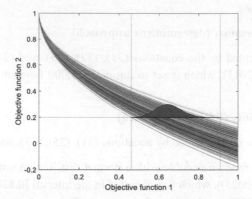

Figure 11. Probability density function of f1 when value of f2 ≈ 0.2

Assume that the decision maker is interested in the values of the objective function 2 in the interval $0.2 \leq f_2 \leq 0.6$ (see Step 7 of ROPAR). To illustrate the analysis of the Pareto fronts, two values of the objective function 2, say 0.2 and 0.6, are picked. Their corresponding PDFs are shown in Figure 11 and Figure 12. Comparing these two PDFs, one can say that in the considered interval, the solutions with the values of $f_2 \approx 0.6$ are the ones with the minimum variability of f_1.

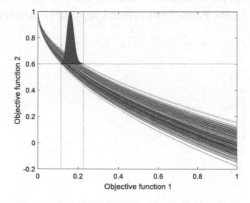

Figure 12. Probability density function of f1 at f2 ≈ 0.6

Steps 8-10 are not carried out in this case due to the fact that the optimisation algorithm finds the ideal (almost exact) solutions for every realization of this random benchmark problem. In other words, these solutions are basically the same and there is no reason to use steps 8-10 of the algorithm which has the purpose of finding the most robust solution among them. The ideal solution of this benchmark problem is when $x_i=0$ for $i=2,3,...,11$.

Comparing the solutions by the deterministic and ROPAR approaches

After carrying out deterministic optimisation of ZDT1, there is no more additional information to help to decide which solution (among the 50 solutions forming the Pareto front) to choose. However, if uncertainty in the benchmark function is assumed, and the robust optimisation approach is followed, one would have more information enabling one of the solutions to be selected. As shown in Figure 11 and Figure 12, the robustness of the solution has a direct relation to the value of the objective function 2: the higher value of the objective function 2, the more robust is the solution. Given the fact that it was assumed that the decision makers are interested in the solutions within the interval $0.2 \leq f_2 \leq 0.6$, the most robust solutions are those corresponding to $f_2 \approx 0.6$.

3.10.4 Conclusions

The simplicity of this problem allows to see clearly the working of ROPAR, and how the uncertainty affects the distribution of optimum solutions. This propagation of the uncertainty from the uncertain parameters to the optimum solutions is one of the contributions of ROPAR to the analysis of robustness. By using ROPAR it is possible to visually analyse the robustness of the solutions influenced by the choice of values of f_2.

3.11 ANALYSIS OF COMPUTATIONAL COMPLEXITY

The optimisation algorithms used in this study employ the multi-objective direct (randomized) search algorithm, e.g. MOO algorithm (NSGAII, AMGA2, etc.). Complexity of such algorithms is measured by the number of objective function evaluations needed, since for real-life problems each such evaluation needs one model run, and it can be computationally demanding (since such model is usually complex, e.g. hydrodynamic). In case of ROPAR MOEA is repeated multiple times following the Monte Carlo framework. Other algorithms used in robust optimisation, like a family of methods generally named here OSOF (optimisation by smoothing the objective function), for each function evaluation require multiple model runs. All other operations in the mentioned algorithms are simple (retrieving the stored values or simple arithmetic operations) and there is no need to take them into account in this analysis. Considering all this, it can be concluded that the computational complexity of algorithms should be measured by the total number of (computational) model runs needed.

3.11.1 ROPAR complexity

As it is stated in Section 3.6.1, ROPAR has four parts. The complexity of each part is analysed to finally determine what the order of the complete algorithm is.

Deterministic optimisation (Part 2) is the generation of Pareto fronts. The number of evaluations of the objective functions to obtain one Pareto front is determined by the criterion used to stop the optimisation. One of such criteria is reaching a prespecified number of

51

evaluations of the objective functions, and it used in this thesis. Let this number be ne, therefore the complexity (number of model runs) of Part 2 is ne.

However deterministic optimisation has to be run multiple times, one for each sample of an uncertain parameter or input vector (this denoted as Part 1 of ROPAR). If the number of samples is np, then the total number of model runs is $np*ne$.

Part 3 (building of an empirical distribution) does not require running the model.

Part 4 is the finding of the robust solution(s). To apply the four robustness metrics belonging to this part, a matrix is required. This matrix is the result of evaluating, for each of the np solutions, the value of its corresponding objective function f_1 for each of the samples np. So the number of operations to calculate the matrix is $np*np$. Furthermore, if analysis is done for several (nl) levels f_2 . Therefore, complexity of Part 4 is $nl*np^2$.

Hence, the total complexity of all ROPAR parts is $ne*np + nl*np^2$.

3.11.2 OSOF complexity

We assume that OSOF (optimisation by smoothing the objective function) uses the same MOO algorithm as ROPAR, and needs ne evaluations of the objective function. However, each evaluation of the objective function requires np executions of the complex computational model (equal to the number of the sampled values or vectors of the uncertain parameter or input). Therefore complexity of OSOF is $ne*np$.

It is clear that ROPAR is more expensive, computationally speaking, because of the additional term $nl*np^2$.

3.11.3 Parallelizing ROPAR and OSOF

Because robust optimisation is computationally demanding, the use of multiple processors would be desirable. One way to achieve higher efficiency of both ROPAR and OSOF is using parallelized version of the MOO algorithm (deterministic optimisation engine), or modifying accordingly it if it is not parallelized.

Efficiency of ROPAR can be further increased: in the Monte Carlo external loop (Steps 1 and 2) each deterministic MOO can be run independently, on separate processors.

Technically, parallelization can be arranged by setting a cluster of several PCs under the Message Passing Interface (MPI), or by using cloud computing services.

4

ROBUST OPTIMISATION OF A SIMPLE STORM DRAINAGE SYSTEM[2]

This chapter presents the application of ROPAR for the design of a small storm drainage network. The problem consists of finding the pipe sizes of the drainage network taking into account two objective functions and one source of uncertainty. The ROPAR algorithm is applied using NSGAII as optimisation engine. The last part of the chapter compares the robustness of a ROPAR solution with the robustness of a deterministic solution.

---

[2] The results of this chapter were published in the following paper:

Marquez-Calvo O. O. & Solomatine D. P. 2019 Approach to robust multi-objective optimization and probabilistic analysis: the ROPAR algorithm. Journal of Hydroinformatics 21, 427-440 doi:10.2166/hydro.2019.095.

4.1 PROBLEM STATEMENT

4.1.1 General problem statement

A storm drainage network is designed to drain excess water from an urban area. To this end, a storm design with a particular return period is traditionally used to decide on the dimensions of the pipes such that the potential damages caused by flooding are minimised. The return period of a storm design typically varies from 2 to 25 years (Akan and Houghtalen 2003).

However, rainfall is a random variable and such uncertainty is not always taken explicitly into account in design practice. Indeed, as rainfall design data is based on historical information, it may be of limited quality. For example, it may be based on unreliable measures, have insufficient historical records. In addition, rainfall patterns may exhibit recent changes that are not captured by historical information (Milly et al. 2008).

Other sources of uncertainty influencing the design include changes in the roughness of the pipes due to their age and use, changes in the land use impacting the imperviousness of the basin, among others.

Robust optimisation is a technique that can help finding the best design of a drainage system within the available budget, taking into account the aforementioned uncertain parameters, which are typically considered as static when the drainage network is designed under a deterministic approach.

This general problem considered in this chapter is optimisation of a simple drainage network, and in the following two chapters, more complexity is added. Formulation here is deliberately simplistic to show the essence of the framework on simple cases. For example, this and the next two chapters are considering pipe diameters as the only variable taken into account when replacing pipes. This is, of course, a simplification, since in reality other variables have to be considered. For example, pipe slopes, which have an impact on the flow velocity leading to erosion problems if they are too high or sedimentation if they are too low, excavation costs, etc. If a real-life case is considered, more factors related to engineering design, additional constraints and additional variables can be added to the main system model and to formulations of objective functions and constraints, however, this will not require changes to the optimisation algorithm itself.

Regarding *flooding volume*, it is determined by running the SWMM modelling software (Rossman 2010). *Flooding volume* is representing the amount of water on the streets, which is used as a surrogate for damages. That is the reason explaining its minimisation. Modelling the water on the streets can be used as input to assess damages. There are works that use directly the volume of water on the streets to carry out risk assessment (Ashley et al. 2005), and there are other works where the depth of water on the streets is used to identify vulnerable locations (van Dijk et al. 2014). That kind of risk assessment is beyond the reach of this thesis.

4.1.2 Particular problem statement

For this experiment, a model of a simple storm drainage pipe network with 11 pipes is used (see Figure 13) (this is a simplified network in a Latin American town). The network has a fixed layout. The decision variables are the diameters of pipes. The mathematical formulation of this problem is

$$\min_{D} Construction\ cost = C_D \cdot L + C_F \qquad (36)$$

$$\min_{D} Flooding\ volume(D, P) \qquad (37)$$

where D is the vector $(d_1, d_2, ..., d_{11})$ representing the diameter of every pipe in the network, C_D is the vector $(c_1, c_2, ..., c_{11})$ representing the cost per length unit of every pipe depending on its diameter, C_F is the fixed cost of the project, L is the vector $(l_1, l_2, ..., l_{11})$ representing the length of every pipe, and P is the amount of precipitation in the basin. In the remainder of this thesis, cost and construction cost are used interchangeably.

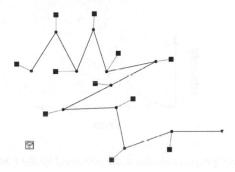

Figure 13. Layout of the storm drainage network with 11 pipes (exported from SWMM software), where lines represent pipes; circles – junctions; squares – subcatchments; and triangle – outfall

4.2 EXPERIMENTAL SETUP

This case is solved deterministically (using NSGAII) and robustly (using ROPAR AD2). To assess if indeed ROPAR effectively finds robust solutions, the deterministic MOO is carried out. These deterministic optimum solutions are used as a baseline to compare them with the solutions found using the robust optimisation approach, using the robustness metrics presented in the methodology.

Design without uncertainty (deterministic approach)

To find the deterministic solutions, the intensity of the design rainfall corresponding to the geographical location of the network is taken into account. The design return period for storm drainage typically varies from 2 to 25 years (Akan and Houghtalen 2003), and for this case 20

55

years was chosen. For convenience, this intensity of the design rainfall is named i_{base} which is also used in the next section.

Two other parameters of the rainfall, duration and pattern, are considered to be fixed, regardless of the intensity (which is, of course, a simplification). With respect to the duration, in the conducted experiments the duration of the rainfall event is taken to be 3 hours. With respect to the pattern of the rainfall, a synthetic hyetograph for design storm was used (Butler and Davies 2004; Haestad and Durrans 2007). The hyetograph resembles the one used by the Chicago method (Keifer and Chu 1957) (see Figure 14). Despite this type of rainfall profile tends to overestimate peak flows (Marsalek and Watt 1984; Alfieri et al. 2008), here it is used because the intention is to design a robust network able to cope with severe events (Fortunato et al. 2014).

The optimiser used is NSGAII. The optimisation is configured to perform 10,000 function evaluations (i.e. runs of the SWMM model).

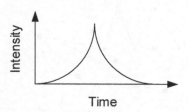

Figure 14. Hyetograph resembling the one used by the Chicago method

Design with uncertainty (ROPAR approach)

For this case, the only uncertain parameter considered is the rainfall. Specifically the intensity of the rainfall is considered to be random (i_{random}). It is assumed that i_{random} follows a normal distribution with a standard deviation of 7% of the mean (i.e. $i_{random} \sim i_{base}*N(1, 0.0049)$), where i_{base} is the intensity of the design rainfall which was defined in the previous section.

The rest of the parameters defining the rainfall were considered not changing in the same fashion as they were described in the previous section.

Information describing the main aspects of the network is as follows. Every pipe has the same data: length of 100 m; slope of 0.5%; Manning's roughness coefficient of 0.02. The areas (in *ha*) of the subcatchments (from upstream to downstream) are: 5.617, 4.441, 4.441, 4.441, 4.089, 4.089, 6.489, 6.489, 3.245, 3.245, and 3.389. The mean value of the accumulated rainfall in 3 hours is 360 mm. The system is configured to consider that in the event of flooding, the excess volume is stored above the junction, in a ponded fashion, and is reintroduced into the system as capacity permits.

The number of samples is determined by using Equation (27). The values for confidence level, level of precision and estimated variability of population are assumed to be 99.90%, 5% and 50%, respectively.

Based on Equation (27), the number of samples is estimated to be 949 which was for the sake of simplicity rounded to 1,000, so that the rainfall intensity i_{random} was sampled 1,000 times. As mentioned above, i_{random} is the result of the multiplication of i_{base} with a random number obtained from the normal distribution N (1, 0.0049). The 1,000 samples fell into the interval [0.7738, 1.2505]

The large sample size (1,000) allows the four objectives to be achieved: first, to reach a high confidence level; second, to have a high level of precision; third, to consider the maximum heterogeneity in the population; fourth, to capture those extreme values that have the potential to be the most troublesome. On the other hand, if a high confidence level or a high level of precision is not needed, or the population does not have a high heterogeneity, then the number of samples can be set to a lower value. This number can be also made lower if the computational load is prohibitively high.

In the case of ROPAR, for each of the sampled rainfall intensities the storm drainage network is optimised using MOO NSGA-II (Deb et al. 2002), resulting in 1,000 Pareto fronts, one front for each sampled rainfall intensity. In every optimisation, 10,000 function evaluations (i.e. runs of the SWMM model needed to estimate flood volume) are used.

Comparing deterministic and robust solutions

To compare the deterministic solution with the robust solutions, the criteria used to determine the robustness in ROPAR are also used to measure the robustness of the deterministic solution. These criteria are specified in Equation (21) and Equation (22).

4.3 RESULTS AND DISCUSSION

The problem of designing a storm drainage network (i.e. the determination of the combination of pipe diameters) is considered, with the two objective functions to minimise: construction cost and flood volume.

Optimising the design without uncertainty (deterministic approach)

Assuming fixed rainfall, the optimum deterministic designs of the storm drainage network are found and shown in the Pareto front in Figure 15. It can be seen that the cost of the solutions is in the range from 1,000 to 4,000 TMU (Thousands of Monetary Units) and flooding volume of these solutions ranges from 0 to 24 MLW (millions of litres of water). Using the Pareto set aids the decision maker in the final choice of the single solution for implementation. For example, if a decision maker considers two possible costs, 1,500 and 3,000 TMU, the corresponding flooding volumes given by the optimal solutions (networks) would be 11.2 and 1.7 MLW, correspondingly. However, the single Pareto set does not give the possibility to say anything about the robustness of these solutions against rainfall uncertainty.

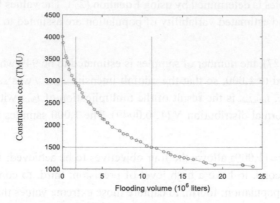

Figure 15. Pareto front representing the solutions of the deterministic optimum designs of the storm sewerage network

Optimising the design under uncertainty (ROPAR approach)

In the ROPAR algorithm, the objective function f_2 is associated with the construction cost (Step 4). Applying ROPAR steps 1, 2, and 3, a set of 1,000 Pareto fronts is generated (Figure 16). In steps 4, 5, and 6, the variability of the flooding volume is analysed for the two construction costs: 1,500 and 3,000 TMU - see Figure 16 and Figure 17, respectively. As in the case of deterministic optimisation, it is easy to see (and it is in a way obvious) that low investment leads to more flooding.

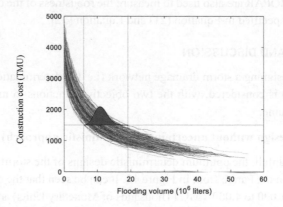

Figure 16. Probability density function of the flooding volume at "Construction cost" ≈ 1500 TMU

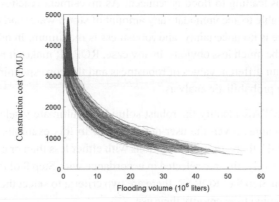

Figure 17. Probability density function of the flooding volume at "Construction cost" ≈ 3000 TMU

However, it is now possible to see also extra patterns: for example, the lower the network costs, the bigger the variability in the network performance. Despite the relatively small uncertainty of the rainfall that is considered (the standard deviation is equal to 7% of the mean), it is interesting to see the differences in the network performance when less and less is invested in the construction of the network.

Let us assume that at Step 7 of the ROPAR algorithm, the decision maker chooses to build a network with a cost of 3,000 TMU. The decision could be based on the fact that although the 3,000-TMU solution is 100% more expensive than the 1,500-TMU solution, the latter leads to more flooding, but also is 350% more variable (measuring variability as the range, i.e. the width of the base of the PDF). The range of the 1,500-TMU solution is 14 MLW; the range of the 3,000-TMU solution is 4 MLW; and the ratio of 14 MLW to 4 MLW is 3.5, or 350%. In other words, the 3,000-TMU network has only 29% of the variability of the flooding volume of the 1,500-TMU network, or it can be said that in these terms it is 3.5 times more robust against uncertainty in rainfall (if one interprets the base of the mentioned PDF as a measure of robustness, but there could be other measures as well). If robustness is seen as an important factor, then a decision maker may find this to be a good additional justification to build a more expensive network - in this case, a 100% more expensive design increases robustness (decreases variability) by approximately 3.5 times, and of course leads to less flooding.

Yet another piece of information that can be deduced from these plots is the relation of maximum flooding volume (MFV) to cost under various scenarios. For example, MFV for the network costing 3,000 TMU is 4 MLW, and MFV for the network of 1,500 TMU is 19 MLW, which is almost 5 times higher.

It should be noted that an increase in robustness for more expensive designs from the engineering point of view is more or less expected: if one invests more in larger pipes, the number of solutions leading to flood is reduced. As investment reaches approx. 4,750 TMU, the pipes would be able to accommodate any amount of storm water and no rainfall would lead to flooding, so there is no uncertainty, and robustness is maximum. In other applications, such a relationship may be much less obvious. In any case, ROPAR makes it possible to identify the sets of solutions with different values of robustness and the corresponding ranges of objective functions, enabling probabilistic analysis.

Steps 8 and 9 of ROPAR identify the robust solution(s), ultimately selecting the most robust of the possible solutions, given the user-defined cost. In this example, for each of the costs 1,500 and 3,000 TMU, there are 1,000 solutions with either less than or equal cost of 1,500 and 3,000 TMU, respectively, that are selected in accordance with Step 8 of the ROPAR algorithm. Next, as specified in Step 9 of ROPAR, there are two criteria to select the most robust solution. Both criteria are applied to exemplify their use.

To apply *Criterion 1* of Step 9 (minimising expected value), for every solution the average flood volume (AFV) across the 1,000 rainfall samples is calculated, and the solution corresponding to the minimum AFV is picked. Figure 18(a) and Figure 18(b), corresponding to costs of 1,500 and 3,000, respectively, show the solutions with minimum AFV marked with a red X.

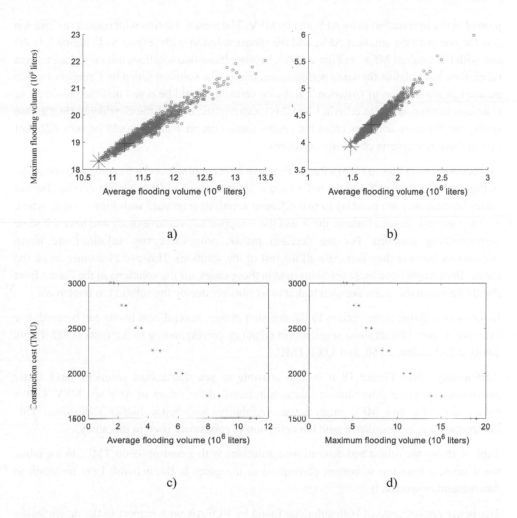

Figure 18. Comparison of robustness of solutions: deterministic (black +), minimum AFV (red X), and minimum MFV (red +); for costs 1500 (a), 3000 (b), and those in-between (c and d)

The procedure described in the previous paragraph is also used with *Criterion 2* of Step 9 of ROPAR (minimising worst case), and here the solution with the minimum MFV is picked. Figure 18(a) and Figure 18(b) corresponding to costs of 1,500 and 3,000, respectively, present the solutions with minimum MFV marked with a red +.

Comparing solutions by deterministic and ROPAR approaches

To see a more general relationship among the 1,001 solutions (i.e. 1,000 from the robust optimisation plus 1 from the deterministic optimisation), in Figure 18(a,b) every solution is

61

plotted in the intersection of its AFV and its MFV. The robust solution with respect to Criterion 1 is the one with the smallest AFV, and the robust solution with respect to Criterion 2 is the one with the smallest MFV. In Figure 18(b), because these two solutions are very close to each other, they have almost the same performance, that is, the solution found by Criterion 1 could be used as the solution of Criterion 2 and vice versa. It should be noted that the proximity of solutions corresponding to criteria 1 and 2 is characteristic only to the considered simple case study, but for more complex cases the results using criteria 1 and 2 could be very different (albeit these two criteria obviously correlate).

Furthermore, from Figure 18(b) a Pareto front can be recognized, and it has only two solutions. In Figure 18(b), the red X and the red + form a 'blurred asterisk' (they almost overlap, but not quite) because they are pointing to two different solutions (compare with Figure 18(a), where a 'sharp asterisk' is seen because the X and the + happen to overlap exactly and have the same corresponding solution). For the decision maker, only these two solutions are worth considering because they dominate all the rest of the solutions. However, for other cases, the Pareto front might have more solutions, and in those cases, all the solutions in the Pareto front should be presented to the decision maker to let him/her decide the solution to implement.

In terms of offering more options to the decision maker, several cost levels can be used. For example, Figure 18(c,d) show seven robust solutions corresponding to the costs 1,500, 1,750, 2,000, 2,250, 2,500, 2,750, and 3,000 TMU.

Additionally, from Figure 18 it is also possible to see that robust solutions have better performance than the deterministic solutions in terms of the values of AFV and MFV. Lower values for AFV and MFV mean lower variability, and hence higher robustness. This interpretation is in accordance with the definition of robustness used in this study.

Table 4 shows the robust and deterministic solutions with a cost of 3,000 TMU. In the table, the diameters from top to bottom correspond to the pipes in the network from upstream to downstream, respectively.

The better performance of both solutions found by ROPAR with respect to the deterministic solution, which is shown in Figure 18(b), can be explained mostly by the difference in diameters of the pipes 5, 6, and 7. The cross sections of these pipes of the deterministic solution are 85%, 92% and 81% of those defined by Criterion 1. Regarding Criterion 2, they are 87%, 95% and 84%, respectively.

Table 4. Solution with "Construction cost" ≈ 3000 TMU found using a deterministic and a robust approach (Criterion 1 and Criterion 2)

	Deterministic	Robustness Criterion 1	Robustness Criterion 2
Pipe Identifier	Diameter (m)	Diameter (m)	Diameter (m)
1	0.81	0.76	0.79
2	1.14	1.15	1.09
3	1.59	1.37	1.36
4	1.64	1.69	1.61
5	1.64	1.78	1.76
6	1.89	1.97	1.94
7	1.89	2.10	2.06
8	2.20	2.19	2.10
9	2.28	2.24	2.34
10	2.28	2.24	2.34
11	2.28	2.28	2.34

4.4 CONCLUSIONS

ROPAR approach uses statistical analysis enhanced by the graphical representation. This combination eases the process of decision making using the graphs as a guide. This graphical part allows to see the propagation of the uncertainty to the optimum solutions. Once a range of solutions is selected, the solutions are analysed using statistical analysis to verify their performance.

The graphical representation is carried out by steps 4, 5, and 6 of ROPAR. These steps allow to see how the uncertainty of the parameters is propagated to the solutions as can be seen in Figure 16 and Figure 17. The uncertainty has the effect of increasing the variability of the solutions set. This variability can be seen as the width of each PDF of the figures just mentioned.

The wider the PDF the more variable are the solutions. Moreover, with more variability the solutions are less robust.

By using the statistical analysis, the robust optimum solution found by ROPAR is compared with the deterministic optimum solution. From this comparison, it is possible to see how far (or close) the ROPAR solution(s) move(s) away from the optimal deterministic one and how much uncertainty influences this deviation.

5

ROBUST OPTIMISATION OF TWO LARGER STORM DRAINAGE SYSTEMS

This chapter describes the application of ROPAR for the design of two complex storm drainage networks. The design problem includes two objective functions and one source of uncertainty. In this chapter the optimisation engine that works along with ROPAR is AMGA2. The chapter finalises with a comparison of the robustness of a ROPAR solution with the robustness of an OSOF solution.

5.1 PROBLEM STATEMENT

The general context of this problem is described in Section 4.1.1. For that reason, it is not repeated here.

The multi-objective optimisation problem in the context of urban drainage network design is posed as follows:

$$\min_{D} Flood(D, R) \tag{38}$$

$$\min_{D} Cost(D) = \min_{D} C_D \cdot L \tag{39}$$

$$C_D, D, L \in \mathbb{R}^p$$

$$Cost, Flood \in \mathbb{R}$$

$$R \in \mathbb{R}^h$$

where D is a vector representing the decision variables which are the diameters of the pipes. R is a vector representing the hyetograph. R could be uncertain or deterministic, requiring a robust or deterministic optimisation, respectively. h is the number of discrete time steps of the hyetograph. *Flood* is the total flood volume and can be calculated using the modelling software SWMM (Rossman 2010). It depends on both D and R. *Cost* is the construction cost of the network. C_D is a vector representing the cost per length unit depending on the diameters D. L is a vector of the lengths of every pipe in the network. p is the number of pipes. $\mathbb{R}$ is the set of real numbers.

This study aims at finding a robust design of a storm drainage network, taking into account only one variable, uncertainty in rainfall (i.e. R). The need to account for this type of uncertainty stems from the fact that using one specific value of design rainfall (for example the one with the return period of 20 years) assumes rainfall stationarity, which is not advisable to use anymore (Milly et al. 2008).

5.2 EXPERIMENTAL SETUP

The experimental setup is composed of the following steps:

1. Setting up the MOO process.
2. Defining the uncertainty used by both RMOO algorithms.
3. Optimising the design without uncertainty (deterministic approach).
4. Optimising the design with uncertainty (Optimisation by Smoothing the Objective Function (OSOF))
5. Optimising the design with uncertainty ('ROPAR AD4' approach).
6. Comparing solutions by all three approaches.

Setting up the MOO process

To carry out robust optimisation of multiple objectives, any multi-objective algorithm can be used; in this study AMGA2 (Tiwari et al. 2011) was employed which was shown to be very efficient and superseding in speed, e.g. NSGA-II. Even so, computational complexity of both OSOF and ROPAR remains high.

For OSOF, the number of evaluations of the objective function was set to be limited to 11,000. The objective function of OSOF is f_μ, and to calculate it, a large number of evaluations of f is needed (1,000 were used), so that in total 11×10^6 evaluations of f would be required. In case of ROPAR, if, say, 1,000 samples are generated, the same number of individual optimisations is needed, each requiring a significant number of the objective function evaluations (a limit of 10,000 was used). Additionally, ROPAR requires 10^6 evaluations of the objective function after the optimisation (see Step 9 of ROPAR). Thus, in total, the number of evaluations of f would be 11×10^6, i.e. it is made equal to the number of evaluations by OSOF.

For complex problems, the number of required evaluations can be made lower, but in any case, it is clear that robust optimisation is CPU intensive. To reduce execution time, an array of PCs was used. The computations were arranged in such a way, that for each sample generated by ROPAR, each AMGA2 optimisation is assigned to one CPU, so no modification of AMGA2 code to parallelize it was required. However, for the case of OSOF, to reduce execution time, AMGA2 required modification to make it distributed-aware; this allowed for employing the MPI (Message Passing Interface) to run computations across several PCs on the network. An option could be using parallelization solutions offered by commercial providers of cloud computing.

In contrast, the computational demand of the deterministic approach was considerably less for obvious reasons: just one sample of rainfall was used instead of the 1,000 samples used by OSOF and ROPAR. AMGA2 was executed with the limit of 10,000 evaluations of the objective function f.

Defining the uncertainty used by both RMOO algorithms

As it is mentioned in Section 5.1, rainfall R is considered to be uncertain. Although the rainfall itself has several parameters such as the profile, the duration and the intensity, for this study only the intensity of the rainfall is considered as an uncertain parameter. It is of course possible to use more comprehensive characterisation of rainfall uncertainty, e.g. following Yazdi et al. (2014) who used joint probability distribution density of duration and intensity, but in this case accurate data and additional statistical analysis for estimating the joint distributions would be needed.

To introduce randomness, the return period T is considered to be a random variable with a half normal distribution (i.e. Equation (40)); this ensures progressive reduction of the event occurrence probability with the return period increase. This distribution has a mean of 46.48 and a standard deviation of 34.36.

$$f(T) = \frac{\sqrt{2}}{\sigma\sqrt{\pi}} e^{-\frac{1}{2}\left(\frac{T-\mu}{\sigma}\right)^2} \; for \; T \geq 1, \mu = 1, \sigma = 57 \tag{40}$$

One thousand samples of T are generated from the probability distribution described by Equation (40); this resulted in the sample range [1.00, 204.97].

To estimate the intensity of the rainfall the IDF (intensity-duration-frequency), the equation presented by Patra (2008) is used. The parameters k, m, and n of IDF Equation (41) are obtained from the historical rainfall information. The IDF equation has two variable parameters, the duration of the rainfall d and the return period T of the event. Duration of the rainfall event d is set to be one hour.

$$i = \frac{k \, T^m}{d^n} \tag{41}$$

For every value of T, the intensity i is calculated using Equation (41), and then converted to a vector R by distributing i according to the proportions in the rainfall profile of Figure 19(c) or Figure 19(d), using a function C. The function of conversion C can be expressed as:

$$C(i) = (i \int_{0}^{\frac{1}{h}} H(t)dt, i \int_{\frac{1}{h}}^{\frac{2}{h}} H(t)dt, \dots, i \int_{\frac{h-1}{h}}^{1} H(t)dt) \tag{42}$$

$$C: \mathbb{R} \rightarrow \mathbb{R}^h$$

where $H(t)$ represents the hyetograph, and h is the number of discrete time steps of the hyetograph.

For the case of the deterministic optimisation, the process just described is also applied, however just one value of return period (i.e. T) is used. To design a storm drainage network, return periods from 2 to 25 years are typically used (Akan and Houghtalen 2003). For the deterministic optimisation, a return period of 20 years was selected (i.e. T=20). This T=20 is approximately in the first quartile of the probability distribution represented by Equation (40).

Optimising the design without uncertainty (deterministic approach)

In order to have a baseline of comparison, a deterministic multi-objective optimisation problem (the one described in Section 5.1) is solved. However, in contrast with the formulation for both OSOF and ROPAR, parameter R is considered to be fixed.

Optimising the design with uncertainty (Optimisation by Smoothing the Objective Function (OSOF))

The methods following the OSOF approach optimise the smoothed version(s) of the original objective functions rather than the objective functions themselves. Smoothing can be achieved by applying some filtering function, e.g. by simple averaging: for each individual (i.e. vector, or point) in the population, a set (sample) of points in the vicinity of this point is considered (set V), for each individual in V the objective function f values are calculated, and then their average f_μ is found. So instead of minimising f, the minimised function is f_μ. Mathematically this can be represented as follows:

$$\min_{x} f_\mu (x, u) \tag{43}$$

$$f_\mu(x, u) = \frac{1}{|V|} \sum_{x, u_i \in V} f(x, u_i) \tag{44}$$

where f is the vector of objective functions, f_μ is the vector of the averages (referred as 'smoothing' in the introduction) of f, V is a set containing the points in the vicinity of (x, u), x is a vector of decision variables, u is a vector of the input variables with uncertainty (randomness), and u_i is the i_{th} instantiation of the uncertain input variable in the neighbourhood. Various, but quite similar, versions of this method of incorporating uncertainty in the formulation of an optimisation problem are employed by a number of authors, e.g. (Jin and Sendhoff 2003; Fieldsend and Everson 2005; Kapelan et al. 2005; Savic 2005; Deb and Gupta 2006; Gaspar-Cunha and Covas 2008; Kuzmin 2009; Kang and Lansey 2012; Zeferino et al. 2012; Vojinovic et al. 2014).

Optimising the design with uncertainty ('ROPAR AD 4' approach)

In this section, 'ROPAR AD4' is the version of the algorithm used (see Section 3.6). 'ROPAR AD4' has four criteria characterising various aspects of robustness of each solution (i.e. Equation (20)). To ease the finding of the most robust solutions, Equation (26) is used.

Comparing solutions by all three approaches

Although OSOF and ROPAR handle robustness somewhat differently, their objective comparison is possible since the comparison is based on 'widely accepted' robustness metrics.

Additionally, using these robustness metrics, the robustness of the deterministic solutions is determined and then compared with the robustness of OSOF and ROPAR solutions.

5.3 CASE STUDIES

Two case studies are considered, one with 100 pipes and one with 59 pipes.

5.3.1 Network with 100 pipes

First the network is described and then the uncertainty taken into account is detailed.

This is an artificial network with 100 conduits (referred to as Net100, Figure 19(a)). For the rainfall records, those for the city of Guachochi, Chihuahua, Mexico were taken into account.

69

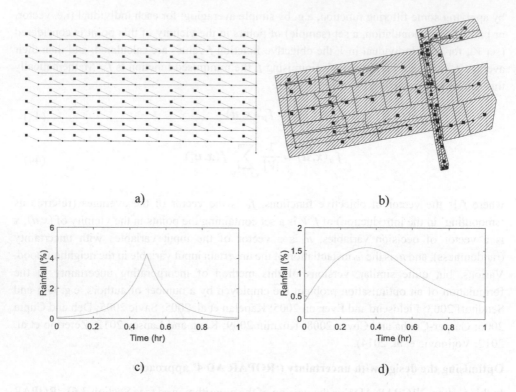

Figure 19. Graphical representation by SWMM (where lines represent conduits; circles - junctions; squares - subcatchments; and triangle – outfall) of the network Net100(a) and its corresponding hyetograph(c), and the same for the network Net59(b) and its hyetograph(d)

Uncertainty source

The uncertainty of the rainfall is modelled using the method described in Section 5.2. From the rainfall records the parameters k, m, and n of IDF Equation (41) are derived. As in the previous case study considered, the rainfall profile (Figure 19(c)) mainly follows the profile used in the Chicago method (Keifer and Chu 1957). This hyetograph represents the function $H(t)$ used in Equation (42). Please see further comments on the choice of hyetograph and the relevant references in Section 4.2.

5.3.2 Network with 59 pipes

This case study (Arachchige-Don 2015) is an approximation of a section of the storm drainage system of Colombo, Sri Lanka, covering the area of 140,300 m^2, and is comprised of 59 conduits (see Figure 19(b)).

Uncertainty source

The uncertainty modelling for this case is similar to the one for Net100. The parameters k, m, and n of IDF Equation (41), and the rainfall profile are based on the rainfall data for Colombo. The hyetograph shown in Figure 19(d) is obtained by applying the method presented by Huff (1990) but assuming the one-hour rainfalls of Colombo. This hyetograph corresponds to the function $H(t)$ of Equation (42). The derivation of duration and hyetograph of the design rainfall for this case is detailed in the next section.

Derivation of the duration and hyetograph of the design rainfall

To find the parameters of the design rainfall to carry out these experiments, historical information of the rainfall of Colombo collected every 15 minutes from 1980 to 2010 is used.

The design duration is found using the recommendation of Klotz et al. (2007). This recommendation is to use the typical storm duration in the area in consideration. To find the typical storm duration, the analysis by Huff (1990) is used. His analysis begins by finding storms from the historical information. He uses six hours as the time between storms, although in this study, one hour is the time to distinguish one storm from the following. Next, the storms are grouped according to their duration (see Table 5 and Figure 20). It is possible to see that storms with a duration of one hour are the most common, therefore one hour is taken as the design duration.

Table 5. Frequency of the storm according to its duration

Range duration (hours)	Frequency
$0 <$ duration ≤ 1	1563
$1 <$ duration ≤ 2	521
$2 <$ duration ≤ 3	252
$3 <$ duration ≤ 4	149
$4 <$ duration ≤ 5	94
$5 <$ duration ≤ 6	41
$6 <$ duration ≤ 7	29
$7 <$ duration ≤ 8	13
$8 <$ duration ≤ 9	6

Range duration (hours)	Frequency
9 < duration ≤ 10	4
10 < duration ≤ 11	4
11 < duration ≤ 12	5
12 < duration ≤ 13	1
13 < duration ≤ 14	0
14 < duration ≤ 15	1

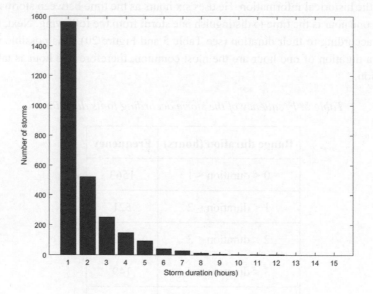

Figure 20. Histogram of storm durations

To define the design hyetograph, another part of the analysis by Huff (1990) is used. In this part the storms are categorized taking into account the quartile where most of the rainfall happens. Those storms where most of the rainfall happens in the first quartile are named Type 1. When it happens in the second quartile, they are named Type 2. The storms Type 3 and Type 4 follow the same logic. The categorization of the storms of Colombo are shown in Table 6. From this table it is possible to see that storms Type 2 are the most frequent. This means that it is more probable that a future rainfall will have its peak in the second quartile. This fact starts to define the 'typical' hyetograph to use in design, because it is possible to see from Table 6 that the storms tend to have their peak in the central part of the hyetograph.

Further, given that the design duration is already defined as one hour, Table 7 is focuses specifically on one-hour duration rainfalls. Here we can also see that these storms tend to have their peak in the central part of the hyetograph. Figure 21 shows the 10th, 50th, and 90th percentile of the time distribution of one-hour rainfalls. We can see that in the median case (50th percentile), 50% of the rainfall happens at approximately 41% of the storm time. From this median case, a hyetograph pattern is derived (see Figure 22). Thus, this is the design hyetograph shown in Figure 19(d).

Table 6. Frequency of the storms according to their category

Category	Frequency
Type 1	298
Type 2	534
Type 3	328
Type 4	159

Table 7. Frequency of the one-hour duration storms according to their category

Category	Frequency
Type 1	53
Type 2	90
Type 3	40
Type 4	16

73

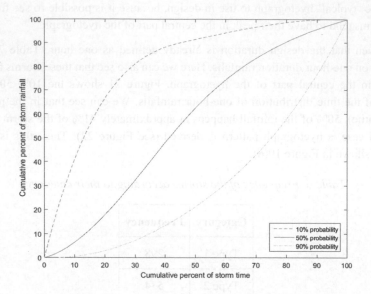

Figure 21. Time distribution of the 10th, 50th (median), and 90th percentile of the one-hour rainfalls

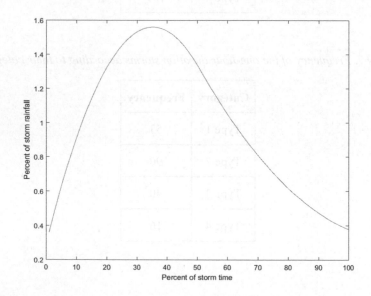

Figure 22. Pattern of the hyetograph derived from the median cumulative time distribution of a one-hour storm

5.4 RESULTS AND DISCUSSION

The analysis begins finding deterministic solutions for both case studies, then robust solutions by both methods OSOF and ROPAR. Finally the robustness of the solutions from all these three methods are compared, to end up discussing the results.

5.4.1 Optimisation without uncertainty (deterministic optimisation)

The result of the deterministic optimisation is the single Pareto front. These Pareto fronts are shown in Figure 23(c) and Figure 23(f) corresponding to Net100 and Net59, respectively, the thick black line, more easily distinguishable on top of the red areas. These Pareto fronts show the trade-off between cost of the network in millions of monetary units (MMU) and the corresponding flood in millions of litres of water (MLW) when that network is undergoing a rainfall with return period of 20 years. Although these Pareto fronts seem to be a continuous line, in fact they are sets of points (i.e. solutions), as it can clearly be seen in Figure 24.

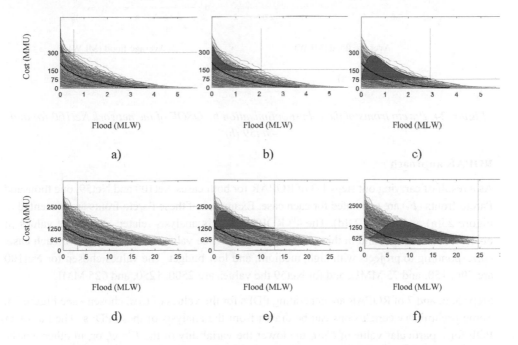

Figure 23. Deterministic and ROPAR optimisations. The deterministic optimisation is the thick black Pareto front more easily distinguishable on top of red areas. ROPAR results are shown up to Step 7. For Net100 the costs equal to 300(a), 150(b), and 75(c) are considered. For Net59 the costs equal to 2500(d), 1250(e), and 625(f) are considered

5.4.2 Optimisation with uncertainty

OSOF approach

As the result of the optimisation using OSOF applied to both case studies, the Pareto fronts are obtained (Figure 24). By definition of OSOF these are robust solutions, and the only piece of information about this robustness is the value of the flood volume, or more precisely, the average flood (see Equation (43)) of each solution when that solution is evaluated for all the samples of rainfall.

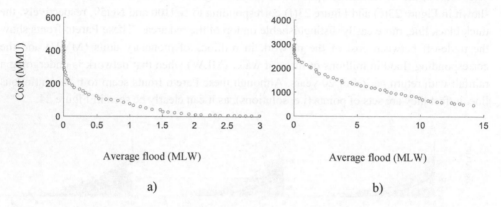

a)   b)

Figure 24. Pareto fronts of the robust optimisation by OSOF of the network Net100 (a) and Net59 (b)

ROPAR approach

As a result of carrying out steps 1-3 of ROPAR for both cases Net100 and Net59, one thousand Pareto fronts (F_r) are generated for each case. Examples of these Pareto fronts are presented in Figure 23(a) and Figure 23(d). The 4th ROPAR step is analysis related to specific values of objective function(s), and in this case this is *Cost*. Three values are considered for each case, corresponding to projects with high, medium, and low budget. The values chosen for Net100 are 300, 150, and 75 MMU, and for Net59 the values are 2500, 1250, and 625 MMU.

Steps 5, 6, and 7 of ROPAR are generating PDFs for the values of *Cost* chosen - see Figure 23. Some preliminary conclusions can be drawn from the analysis of these PDFs. The narrower PDF for a particular value of *Cost,* the lower the variability of the *Flood*, or, in other words, higher the robustness of the solutions. This is verified in following steps of ROPAR.

In Step 8 of ROPAR, all the solutions with *Cost*=*Cost** (e.g. *Cost*=2500) are selected to be included in *S*. However, given that not all (or none of) solutions have an exact value of *Cost**, an interval of *Cost* is used to select the solutions. In this experiment, this interval is [0.88*Cost**,*Cost**] (e.g. [2200,2500]). The number of solutions selected for each case can be seen in the maximum solution index of each case in Figure 25. For example in the Net59 case with *Cost*=2500 (Figure 25(d)), the number of solutions selected is 368 (i.e. |*S*|=368).

In Step 9 of ROPAR, the quadruple defining the *RobustnessIndicator* (i.e. Equation (20)) of every solution *s* in *S* is calculated. To exemplify the calculation of *RobustnessIndicator* (*RI*), the calculation of *Meanf1* (i.e. Equation (21)) is considered. Begin considering that *Meanf1=MeanFlood*. Given a specific solution *s* to evaluate, to calculate *MeanFlood(s)*, it is necessary to calculate *Flood(s,R_k)* where R_k is the *k*-th instantiation of parameter of rainfall with uncertainty. An important piece of information is the range of *k*, which is taking values from 1 to *np* (where *np* is 1,000 given that 1,000 samples of rainfall are considered). Once having the 1,000 values of *Flood*, they are added-up and divided by *np* (i.e. by 1,000). This quotient is the value of *MeanFlood* (*s*). All other elements of the quadruple (*StdDevFlood*, *MaxFlood*, *NonZeroFlood*) forming *RI*, are handled in a similar fashion; the same range of *k* has to be considered.

Once having the *RI* of every solution *s* in *S*, the solution with the lowest value of the *RI* quadruple can be found. To illustrate graphically the process of finding the most robust solution, the value of *NormRobustnessIndicator* (i.e. Equation (26)) is calculated for every *s* in *S*. Then all the values of *NormRobustnessIndicator* (*NRI*) are ordered increasingly and then plotted in Figure 25. The most robust solution is the one with index 1, i.e. the first solution from left to right. This solution leads to the least variability in terms of Equations (20)-(24), and, matches the definition of robustness in Section 2.3.1.

It is worthwhile to point out two extreme cases, considered for Net59. First case, where the difference between the most and the least robust solution is the largest, is when *Cost*=2500 (Figure 25(d)) - here the ratio *NRI* (s_{least})/*NRI* (s_{most}) is 23.8. Second case, where the difference is the smallest, is with *Cost*=625 (Figure 25(f)), where the ratio is equal to 1.4.

5.4.3 Comparison of solutions found by OSOF, ROPAR and deterministic optimisation

Following the procedure defined in Section 5.2, the value of *RI* (x_{OSOF}) is calculated to compare OSOF with ROPAR. To simplify its comparison with all the solutions found by ROPAR, including the most robust solution s_{most}, x_{OSOF} is included in *S*. Once in *S*, x_{OSOF} is named s_{OSOF} and its *NRI* (s_{OSOF}) is evaluated, and then included in Figure 25. To have a numerical comparison between the OSOF solution and the most robust solution of ROPAR, the ratio *NRI*(s_{OSOF})/*NRI*(s_{most}) is calculated. The ratios for the cases in Figure 25 are 2.0(a), 1.5(b), 1.3(c), 7.7(d), 1.4(e), and 1.1(f).

To compare the deterministic solutions with the ROPAR solutions, a procedure similar to the one in the previous paragraph is used. The deterministic solution x_{det} is included in *S*, to be then named s_{det}. The value *NRI*(s_{det}) is calculated and included in Figure 25. As it can be seen in this figure, the deterministic solution appears only in Figure 25(c), Figure 25(e), and Figure 25(f). This is because there were no deterministic solutions in the interval [0.88*Cost**,*Cost**] for the rest of the cases.

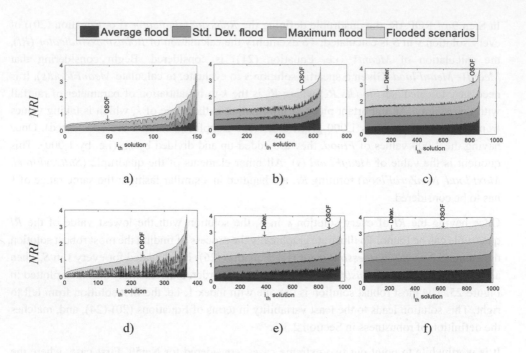

Figure 25. ROPAR results of steps 8 and 9, and comparison with OSOF. Net100 case for Cost equal to 300(a), 150(b), and 75(c). Net59 case for Cost equal to 2500(d), 1250(e), and 625(f). The labels Deter and OSOF are pinpointing the rank of the deterministic and OSOF solutions with respect to all ROPAR solutions. The most robust solution found by ROPAR has index 1

5.4.4 Discussion

Analysing all solutions found by ROPAR for both networks with low budget (cost) (Figure 25 (c) and (f)), it is possible to see that all of them have similar robustness. The four elements of the quadruple of RI have similar values, see Figure 26(i,j,k,l) and Figure 26(u,v,w,x). For example, considering average flood, most of the solutions have almost the same value, see Figure 26(i) and Figure 26(u). Nevertheless, as it is mentioned in the previous section, the ratio $NRI(s_{\text{least}})/NRI(s_{\text{most}})$ for Figure 25(f) is 1.4, which means that the least robust solution leads to 40% more variability with respect to the most robust solution, and this difference is quite considerable, this can be seen in more detail in Figure 26(u,v,w,x). The same ratio for Figure 25(c) is 1.6, which means even higher difference; this greater variance comes from Figure 26(i,j,k,l).

Before further discussion, it is worth making a note about Figure 26. Coinciding with the interpretation of NRI, those solutions in Figure 26 with rank 1 are the most robust because they have the lowest variability, which matches the definition in Section 2.3.1.

Continuing with the discussion, with respect to the solutions with the medium budget, the solutions have approximately the same behaviour as those of low budget, meaning that most of the solutions have similar values of the *RI* quadruple. However here the variability is larger, the ratio $NRI(s_{least})/NRI(s_{most})$ for Figure 25(b) and (f) being 2.4 and 1.7, respectively. The source of this variability can be seen in Figure 26(e,f,g,h) and Figure 26(q,r,s,t).

In contrast, the solutions with high budget are very different. The ratio $NRI(s_{least})/NRI(s_{most})$ for Figure 25(d) is 23.8, meaning 2380% more variability for the least robust solution w.r.t. the most robust one, this very high variability between the least and most robust can be seen in more detail in Figure 26(m,n,o,p). The ratio for Figure 25(a) is 11.8, see also Figure 26(a,b,c,d).

It is now possible to compare the most robust solutions of ROPAR with the solutions from OSOF. The ratio $NRI(s_{OSOF})/NRI(s_{most})$ is used (see Section 5.4.3). The highest ratio is 7.7 corresponding to Figure 25(d), this large variation corresponds basically to the difference in the number of events with flood (see Figure 26(p)). The lowest ratio is 1.1 for Figure 25(f), this small difference between the ROPAR and OSOF solutions can be also seen in Figure 26(u,v,w,x). The ratios 7.7 and 1.1 could be interpreted as 670% and 10% more variability, respectively, if the OSOF solutions is used instead of the most robust solutions of ROPAR.

Interestingly and surprisingly, for the cases where there exist a deterministic solution, these solutions are more robust than the OSOF solutions, and this can be seen in Figure 25(c) and Figure 25(e,f) corresponding to Net100 and Net59, respectively. To discern more clearly how much the deterministic solutions are more robust with respect to OSOF in each criterion of robustness, see Figure 26(i,j,k,l) and Figure 26(q,r,s,t,u,v,x,w) corresponding to Net100 and Net59, respectively. In contrast, ROPAR is capable of finding solutions more robust than the ones found by a deterministic optimisation, as can be seen in the same figures mentioned in this paragraph. It is worth pointing out that in Figure 26(t,x), although the deterministic solutions appear to be better, in reality the solutions from the three methods (i.e. OSOF, ROPAR, and deterministic) have the same value and their position is explained by the inner working of the sorting algorithm ranking the solutions.

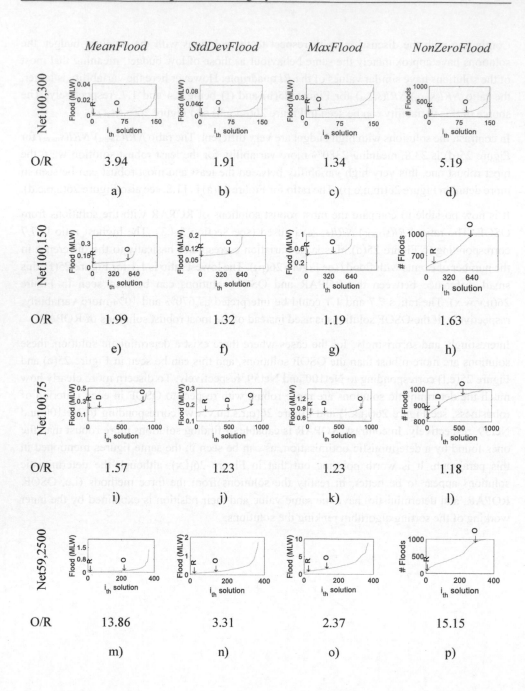

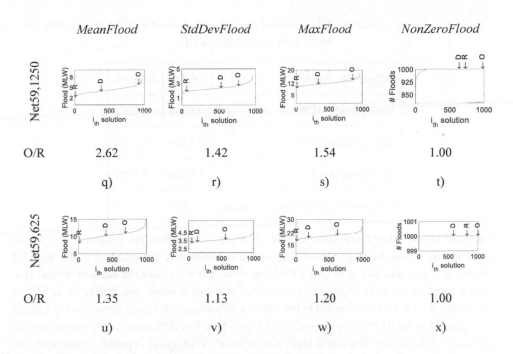

Figure 26. Ranking of the deterministic, OSOF, and ROPAR solutions according to the four criteria of robustness. R=ROPAR solution; D=deterministic solution; O=OSOF solution; O/R=ratio OSOF-metric / ROPAR-metric

To see better what the optimal solutions actually are, the solutions corresponding to the network Net59 with cost 2,500 MMU are analysed (see Table 8). The OSOF solution has 13 times more water overflow (measured in SWMM in volume) on the streets than the ROPAR solution (averaged across the uncertain rainfall). However in the worst case, the OSOF solution has just 1.4 times more water overflow than the ROPAR solution. What it is noticeable is that the reliability (robustness Criterion 4) of the OSOF solution is very low because 894 scenarios out of 1,000 lead to water overflow (*NonZeroFlood*). In contrast, the ROPAR solution has a higher reliability because just for 59 of those 1000 scenarios there is water overflow. This difference in reliability is explained by the fact that the OSOF solution is characterised by the 2.3 times higher standard deviation of overflow volume (*StdDevFlood*) than the ROPAR solution.

81

Table 8. Robustness metrics of the solutions found by OSOF and ROPAR, for the network Net59 with Cost≈2,500 MMU.

Robustness metric	OSOF	ROPAR
MeanFlood (MLW)	0.0864	0.0062
StdDevFlood (MLW)	0.1404	0.0424
MaxFlood (MLW)	1.4440	0.6090
NonZeroFlood (Number of floods)	894	59

Figure 27 shows what these solutions are. In these two figures the thickness of the line represents the Cross Section Area (CSA) of the conduit. It is noticeable that the ROPAR solution is using conduits with larger CSAs in those conduits near the network outfall (i.e. conduit most to the left). Figure 27 (c) shows, for every conduit, the difference in CSAs, specifically the CSA of one conduit in the OSOF solution (i.e. CSA_{OSOF}) minus the CSA of the same conduit in the ROPAR solution (i.e. CSA_{ROPAR}), so this difference can be represented as (CSA_{OSOF} - CSA_{ROPAR}). The line is black if (CSA_{OSOF} - CSA_{ROPAR}) is positive, otherwise, the line is red. The thickness of the line is proportional to one of the following two ratios indicating relative change in CSA: if the line is black, it is proportional to (CSA_{OSOF} - CSA_{ROPAR})/CSA_{OSOF}; if red, to (CSA_{ROPAR} - CSA_{OSOF})/CSA_{ROPAR}. From Figure 27 (c) we can see that the ROPAR solution is using larger pipes from nodes upstream to nodes downstream.

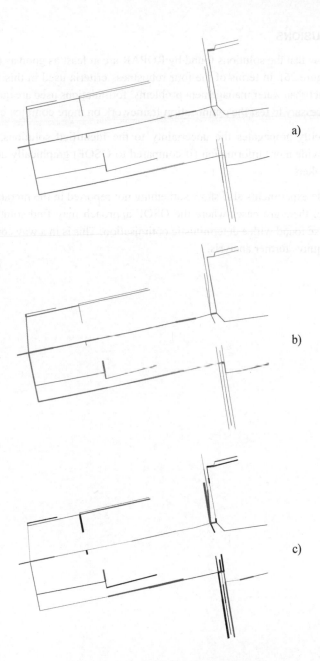

a)

b)

c)

Figure 27. Solutions for the case Net59 with Cost≈2500 MMU, OSOF (a), ROPAR (b) and the differences between these two (c).

83

5.5 CONCLUSIONS

The results show that the solutions found by ROPAR are at least as good as the ones found by OSOF (see Figure 26), in terms of the four robustness criteria used in this thesis. It is worth mentioning that urban water management problems' formulations used are quite simplistic, and it would be necessary to test the optimisation framework on more complex case studies.

ROPAR explicitly propagates the uncertainty to the identified solutions, which makes it possible to provide more information (if compared to OSOF) graphically and analytically to the decision makers.

Moreover, these experiments also show something not reported in the literature, to the best of our knowledge: there are cases where the OSOF approach may find solutions that are less robust than those found with a deterministic optimisation. This is in a way counterintuitive and this finding requires further analysis.

6

ROBUST OPTIMISATION OF A STORM DRAINAGE SYSTEM: MORE OBJECTIVES AND SOURCES OF UNCERTAINTY

In this chapter the design of yet another storm drainage system, of higher complexity, is optimised. Its complexity stems from the three factors not included in previous chapters. First, the design is also considering what is known as Best Management Practices. Second, the problem has three objective functions, and a version of ROPAR that can handle more than two objective functions is employed. Third, the three sources of uncertainty are taken into account. Furthermore, as it has been done in previous chapters, the robust design by ROPAR is compared with the deterministic design by AMGA2.

6.1 PROBLEM STATEMENT

The general context of this problem is described in Section 4.1.1. For that reason, it is not repeated here.

The problem is to find a robust optimum design of a storm drainage system. This design consist in the definition of the pipe diameters and the definition of the Best Management Practices (BMPs) to be implemented. There are three objective functions: first, minimise construction cost; second, minimise floods; third, maximise amount of flood water infiltrated to groundwater. In order to have consistency in the form of the objective functions, the third objective function is reformulated to convert the maximisation problem to the minimisation one: the third objective becomes the minimisation of flood water not infiltrated (*WNI*) to groundwater. The decision variables are the pipe diameters and the type and proportion of BMPs to be implemented in the basin. There are three sources of uncertainty: changes in the stationarity of rainfall, the roughness of the pipes representing their aging, and changes in land use of each subcatchment impacting its imperviousness. The formulation of this problem is as follows:

$$\min_{D,B} Flood(D, B, Rai, Rou, Imp) \tag{45}$$

$$\min_{D,B} Cost(D, B) = \min_{D,B} C_D \cdot L + C_B \cdot B \tag{46}$$

$$\min_{D,B} WNI(D, B, Rai, Rou, Imp) \tag{47}$$

$$C_D, D, L, Rou \in \mathbb{R}^p$$

$$Cost, Flood, WNI \in \mathbb{R}$$

$$Rai \in \mathbb{R}^h$$

$$Imp \in \mathbb{R}^s$$

$$C_B, B \in \mathbb{R}^a$$

where D is a vector representing decision variables which are the pipe diameters. B is a vector representing the proportion of BMPs implemented in the basin, which is also a decision variable. Rai is a vector representing the hyetograph. h is the number of discrete time slots of the hyetograph. Rou is a vector representing the roughness of the pipes. Imp is a vector representing the imperviousness of the subcatchments. $Flood$ and WNI are calculated using the modelling software SWMM (Rossman 2010). WNI is the percentage of the total rainfall in the basin that does not go to groundwater. $Cost$ is the construction cost of the storm drainage system. C_D is a vector representing the cost per length unit depending on the pipe diameters D. L is a vector of the lengths of every pipe in the network. C_B is a vector of the cost of the BMP depending on the BMPs B used. p is the number of pipes. a is the number of areas using BMPs. s is the number of subcatchments in the basin.

Differently from the previous chapters where a symbolic currency was used (i.e. 'Monetary Units'), in this chapter the US Dollar (i.e. USD) is used. This was necessary in order to have a realistic ratio between the cost of a 'pipe project' and the cost of a 'BMP project' given that the optimisation algorithm is deciding how to allocate the budget between these two kinds of projects.

6.2 EXPERIMENTAL SETUP

The main steps in experimental work are as follows:

1. Find optimum designs of the storm drainage network using a deterministic approach.
2. Find optimum designs of the storm drainage network using a robust approach ('ROPAR AD2').
3. Compare the robustness of these two designs.

Next, these steps are explained in more detail.

6.2.1 Designs without uncertainty (deterministic approach)

Similarly to the previous case study, to identify the deterministic design, the AMGA2 (Tiwari et al. 2011) is used. However, different from the previous case study, the deterministic optimisation is carrying out 10^7 evaluations of the objective function instead of 10^4. The intention is to find the best Pareto front using the same number of evaluations of the objective function as those carried out by the robust optimisation. With this in mind and in order to find the best possible solutions within the time frame, one thousand optimisations are executed with different seeds each. These seeds are generated from sampling from the uniform distribution unif(0,1). This is possible because AMGA2 uses a modifiable seed for its pseudo random number generator.

Each of these one thousand optimisations is configured to obtain a Pareto front after 10,000 evaluations of the objective function. At the end of the optimisations, one thousand Pareto fronts are obtained. The solutions of all these Pareto fronts are combined to obtain one global Pareto front, presenting the final result of the deterministic optimisation.

6.2.2 Designs under uncertainty (ROPAR approach)

In Step 3 of ROPAR, the MOO used is AMGA2. Coinciding with the deterministic optimisation, the number of function evaluations to carry out each optimisation is 10,000. The number of samples of the uncertain parameters are 1,000 (defined in step 1). And given that every sample is used in one optimisation, then 1,000 optimisations are carried out. The details of the sampling are explained further in the section describing the case study.

6.2.3 Comparing solutions by deterministic and ROPAR approaches

The solution found by ROPAR (i.e. x_{ROPAR}) has $f_2=f_2{}^*$. The solution with $f_2=f_2{}^*$ is located in the global Pareto front generated by the deterministic optimisation, such solution is named x_{DET}. The robustness of x_{DET} is calculated by determining its 'expected value' and 'worst case' (Criteria 1 and 2 of ROPAR Step 9). Now it is possible to compare the robustness of both solutions, by comparing the respective values of robustness criteria, 'expected value' and 'worst case'.

6.3 CASE STUDY

In this section, the features of the case study are described. It is also described how the parameters *Rai*, *Rou*, and *Imp* (see Section 'Problem statement') are defined for the robust and deterministic optimisations. The case of the robust optimisation is presented first because it eases the explanation of the deterministic optimisation.

6.3.1 Description of the case study

This is an artificial network created for the author of this document, the same as presented in Chapter 5. It has 100 conduits. Its graphical representation is shown in Figure 19(a). It is assumed that this network is located in Guachochi, Chihuahua, Mexico. The real location is needed because the historical rainfall information of this place is used to calculate the design rainfall, and design rainfall is one element used in the optimisation of this network.

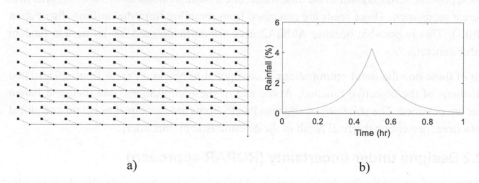

a)   b)

Figure 28. Graphical representation by SWMM of the network (a) and its corresponding hyetograph (b)

The new element in the current case study is consideration of BMPs. This network has 100 subcatchments. Each subcatchment has three kinds of impervious areas: pavement, roof, and common area (area between buildings). These three kinds of impervious areas can be subject to BMPs in the following manner. Pavement can be replaced by permeable pavement. Roof can be made a green roof. Common area can be transformed by adding various infrastructural solutions: a bio-retention cell, rain garden, infiltration trench, permeable pavement, or vegetative swale.

It is considered that maximum 50% of each kind of impervious area (i.e. pavement, roof, and common area) can be treated with BMPs. The amount of this treatment is a decision variable, so that it is defined by the optimisation.

6.3.2 Uncertainty sources: rainfall, pipe age and catchment characteristics

For the case of the robust optimisation, *Rai*, *Rou*, and *Imp* are considered as independent random variables. This section is describing each of these sources of uncertainty.

Let us begin defining *Rai*. It is defined exactly as it is described in Section 6.1. One piece of information that is necessary to complete the definition of *Rai* is the determination of rainfall profile to use in this case study. The profile to be used is the one shown in Figure 28(b). Furthermore, *Rai* is modelled using a half-normal distribution as it is described in Section 5.2, Subsection 'Defining the uncertainty used by both RMOO algorithms'.

The second uncertain parameter is *Rou*. It represents the uncertainty of pipes roughness due to aging. In this network, it is assumed that the pipes are made of concrete. The roughness of the pipe, represented by k_s (mm), varies in the interval [0.06,6.0] depending on its age (Butler and Davies 2004). It is assumed that the roughness has a uniform distribution in this interval, and one thousand values of *Rou* are sampled.

The last parameter with uncertainty is *Imp* – it is representing the change of land use of the subcatchment through the lifetime of the storm drainage system, this change impacts the imperviousness of the subcatchment. *Imp* is a percentage, the percentage of the subcatchment that is impervious. It is assumed that *Imp* follows a normal distribution, and one thousand values that *Imp* takes are obtained by sampling from a normal distribution with mean of 0.70 and standard deviation of 0.05 (i,e. 70% of each subcatchment is assumed to be impervious).

To ensure an even distribution of the sampling of these three random variables (i.e. *Rai*, *Rou*, and *Imp*), a Latin hyper cube sampling is used to generate the 1,000 samples.

For the deterministic optimisation, *Rai*, *Rou*, and *Imp* are constant, fixed at a specific value. *Rai* is defined after considering a specific return period. A design return period for storm drainage typically varies from 2 to 25 years (Akan and Houghtalen 2003), and we set it to 20 years. The value of *Rou* is defined after considering a k_s (mm) equal to 0.20. *Imp* is set to be 70%.

6.4 RESULTS AND DISCUSSION

This section is organized in the following manner: first, the results of the deterministic optimisation; second, the results of the robust optimisation; third, the analysis of these two results.

6.4.1 Optimisation without uncertainty (deterministic approach)

As a result of applying the method explained in Section 6.2.1, a global Pareto front is obtained. Figure 29 shows the two projections of this three-dimensional Pareto front: Cost vs Flood and Cost vs WNI.

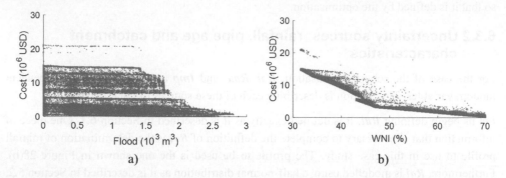

Figure 29. Projections of the three-dimensional Pareto front of the deterministic optimisation: Cost vs Flood (a) and Cost vs WNI (b)

6.4.2 Optimisation with uncertainty (ROPAR approach)

As a result of carrying out steps 1-3 of ROPAR, one thousand Pareto fronts are generated. Examples of these Pareto fronts can be seen in the second plane of Figure 30(a) and Figure 30(d). Step 4 of ROPAR is analysing a specific value of *Cost*. Three values are considered, corresponding to projects with low, medium, and high budget. The values chosen are 3, 6, and 12 (10^6 USD).

Steps 5, 6, and 7 of ROPAR are generating PDFs for the specified values of *Cost* chosen. Specifically, the procedure followed to generate the PDFs is explained in Section 3.8, Consideration 1. The resultant PDFs are shown in Figure 30, and some preliminary conclusions can be drawn. The smaller PDF width for *Cost*=12 with respect to the width of the PDF for *Cost*=3 is an indication of lower variability of the *Flood* or in other words, more robust systems.

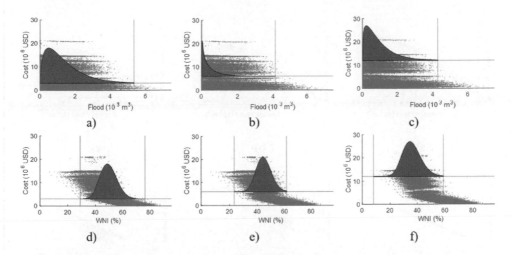

Figure 30. ROPAR results after the first 7 steps for solutions with Cost=3 (a,d), Cost=6 (b,e), and Cost=12 (c,f)

To carry out Step 8 of ROPAR, all those solutions with *Cost=Cost** (e.g. *Cost*=12) are selected to be included in *S*. But given that not all solutions have an exact value of *Cost**, then an interval of the *Cost* have to be considered to select the solutions. That interval is [0.90*Cost**,1.10*Cost**] (e.g. [10.8,13.2]).

Step 9 of ROPAR finds the most robust solutions by applying Criterion 1 and 2. This step can be carried out numerically, however here a visual approach is used. For every solution selected in Step 8, the average and maximum value of flood is calculated, as well as the average and maximum value of WNI. Specifically the procedure followed is described in Section 3.8, Consideration 2. The results of applying this procedure are plotted in Figure 31. Every point in the figure represents a solution. Solutions closer to the origin are more robust because both Criterion 1 and Criterion 2 (which are to be minimised) have low values (in accordance with the definition of robustness of Section 2.3.1).

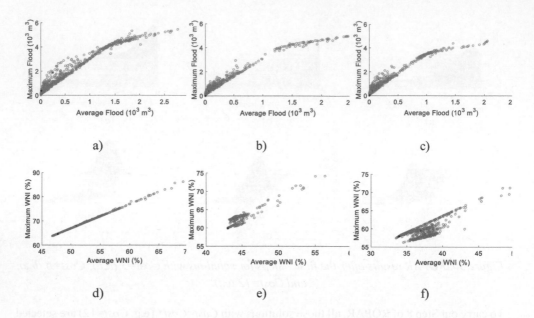

Figure 31. Determining the robustness of all the solutions, including the ROPAR solution (black dot) and deterministic solution (red dot), with Cost=3 (a,d), Cost=6 (b,e), and Cost=12 (c,f)

6.4.3 Comparing solutions by the two approaches and discussion

The solutions with *Cost* of 3, 6, and 12 are selected from the deterministic global Pareto front of Figure 29. In order to select these solutions, the same procedure regarding Step 8 of ROPAR explained in the previous section (i.e. Section 6.4.2) is used. For these selected solutions, the method defined in Section 6.2.3 is applied, i.e. the respective values for the two robustness criteria ('expected value' and 'worst case') are calculated. The solution for each of the selected Costs are plotted in Figure 31 as the red dots. It is worth explaining the meaning of the red dot representing the deterministic solution. Using for this explanation the case of *Cost*=3, the red dot represents the same solution for both Figure 31(a) and Figure 31(d); in this way, we can see its robustness regarding flood and WNI, respectively. In the same fashion, for the case of *Cost*=12, the red dot represents the same solution for both Figure 31(c) and Figure 31(f). We can state the same for the case of *Cost*=6, although for this case the red dot is not visible in Figure 31(e) because the red dot occupies exactly the same place as the black dot.

As it can be seen in Figure 31, ROPAR finds several solutions that are more robust than the solution represented by the red point (those solutions closer to the origin). The next step is to determine one solution that, at the same time, is robust with respect to flood and also with respect to WNI. To choose one solution as the most robust, the procedure explained in Section 3.8, Consideration 3, Option 3.1 is followed. These most robust solutions are represented by the black dots in Figure 31.

It is clear from every graph in Figure 31 that ROPAR finds solutions more robust than the deterministic solutions. Including the graph of Figure 31(e), where the red dot and the black dot are located exactly at the same point. Being in the same location means that the two solutions have exactly the same robustness with respect to WNI; however with respect to flood, the ROPAR solution is better (Figure 31(b)). And given that the black dot represents the same solution in Figure 31(b) and Figure 31(e), the ROPAR solution is more robust as a whole than the deterministic solution.

In this section, *Cost* is the pivotal objective function and the other two objective functions (i.e. *Flood* and *WNI*) are the nonpivotal objective functions. Measuring the variability of these two nonpivotal objective functions results in the determination of the robustness. Every nonpivotal objective function represents a different kind of robustness. It is possible to find a solution that minimises all kind of robustness at the same time. However, it is also possible to find compromised solutions, for example prioritizing one objective function over the others. It is worth saying that this type of information cannot be extracted from OSOF.

As it can be seen, this idea of taking one objective function as the pivot, and taking the rest of the objective functions as nonpivot, can be extended to a problem with n objective functions.

Now, we are going to examine in detail the differences between the solutions identified by deterministic optimisation, and by ROPAR. To exemplify this, we will be considering solutions with $Cost \approx 3*10^6$ USD. Table 9 shows the robustness metrics of these solutions. It can be seen that the deterministic solution has, in average, 3450% more water overflow volume than that of the ROPAR solution. With respect to the maximum volume, the deterministic solution has 370% more than the ROPAR solution. However regarding the WNI, both solutions have similar performance because the deterministic solution has 6% more average WNI that those of the ROPAR solution. Similarly, the maximum WNI of the deterministic solution is just 5% more than the ROPAR solution.

Table 9. Robustness metrics of the deterministic and ROPAR solutions with $Cost \approx 3*10^6$ USD

Robustness metric	Deterministic	ROPAR
Average flood	0.2451	0.0069
Maximum flood	1.8511	0.3975
Average WNI	50.8429	47.7471
Maximum WNI	67.7537	64.6443

The specific configurations of the solutions are represented in Table 10 and Figure 32. Table 10 shows that both solutions selected the same kind of impervious area and the same kind of treatment and approximately the same percentage of treated area. This explains why both solutions have similar performance regarding WNI.

Table 10. Configuration of the BMPs for deterministic and ROPAR solutions with
Cost≈3*10⁶ USD

Impervious area	Treatment	Deterministic Treated area (%)	ROPAR Treated area (%)
Common area	Bio-retention cell	0	0
Common area	Rain garden	0	0
Common area	Infiltration trench	0	0
Common area	Permeable pavement	0	0
Common area	Vegetative swale	41.24	48.34
Pavement	Permeable pavement	0	0
Roof	Green roof	2.24	0

The big difference in performance regarding flood (Table 9) is explained by the differences in pipe diameters (Figure 32). In this figure the thickness of the line represents the Cross Section Area (CSA) of the pipe. It can be seen that the ROPAR solution is more effective in minimising manhole surcharge, explaining its better flood performance. To see more clearly what the changes of one solution with respect to the other are, Figure 32 (c) is presented. This figure is indicating, for every pipe, the difference in CSAs, specifically the CSA of one pipe in the deterministic solution (i.e. CSA_{DET}) minus the CSA of the same pipe in the ROPAR solution (i.e. CSA_{ROPAR}), then this difference can be represented as (CSA_{DET} - CSA_{ROPAR}). The line is black when (CSA_{DET} - CSA_{ROPAR}) is positive. If (CSA_{DET} - CSA_{ROPAR}) is negative, then the line is red. If (CSA_{DET} - CSA_{ROPAR}) is zero, then no line is displayed. That is why the following pipes are not displayed: the first pipe, from left to right, of every series of horizontal pipes nor the first nor the second, from bottom to top, of the series of vertical pipes. In Figure 32 (c), the thickness of the line is proportional to either of these two ratios: if the line is black, (CSA_{DET} - CSA_{ROPAR})/CSA_{DET}; if red, (CSA_{ROPAR} - CSA_{DET})/CSA_{ROPAR}. So that the thickness is an indication of the proportion of change.

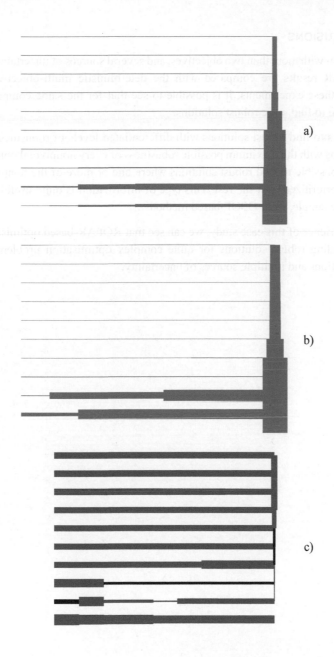

Figure 32. Solutions for the case with Cost≈3*10^6 USD: deterministic (a), ROPAR (b), and the differences between these two (c).

6.5 CONCLUSIONS

Here, a problem with more than two objectives, and several sources of uncertainty is considered. Again, ROPAR results are compared with the deterministic multi-objective optimisation results. From these experiments, it is possible to see that for the same computational effort, ROPAR is able to find more robust solutions.

ROPAR allows to find robust solutions with differentiated levels of robustness. It is possible to find solutions with the maximum possible robustness of every nonpivotal objective function, and it is also possible to find robust solutions where one or more of the nonpivotal objective functions are prioritized over the rest. This type of information is quite useful, and cannot be extracted, for example, from OSOF-based methods.

From the experience of this case study, we can see that ROPAR-based optimisation procedure allows for finding robust solutions for quite complex optimisation problems with several objective functions and multiple sources of uncertainty.

7

ROBUST OPTIMISATION OF WATER QUALITY IN DISTRIBUTION SYSTEMS[3]

This chapter is devoted to solve a particular sub-problem in dealing with water quality in distribution networks, - minimisation of water age in a network, with four examples. The chapter begins with the development of a new two-objective optimisation algorithm more suitable to solve this particular problem than the randomized search (genetic) algorithms used in previous chapters. This new optimisation algorithm is used within ROPAR as the deterministic optimisation engine. Finally an analysis is carried out to reflect on the importance of using robust optimisation instead of deterministic one.

---

[3] The results of this chapter were published in the following papers:

Marquez-Calvo O. O., Quintiliani C., Alfonso L., Di Cristo C., Leopardi A., Solomatine D. P. & de Marinis G. 2018 Robust optimization of valve management to improve water quality in WDNs under demand uncertainty. Urban Water Journal 15, 943–952 doi:10.1080/1573062X.2019.1595673.

Quintiliani C., Marquez-Calvo O. O., Alfonso L., Di Cristo C., Leopardi A., Solomatine D. P. & de Marinis G. 2019 Multi-objective valve management optimization formulations for water quality enhancement in WDNs. Journal of Water Resources Planning and Management 145, 04019061 doi:10.1061/(ASCE)WR.1943-5452.0001133.

7.1 WATER AGE MINIMISATION PROBLEM AND ITS DETERMINISTIC SOLUTION

In this section the water quality problem in distribution systems is introduced, and solved using multi-objective deterministic optimisation. Essence of this optimisation is in finding configuration of the water distribution network to minimise the water age. So far, for deterministic optimisation only evolutionary MOOs have been used, however specifics of the problem at hand required development of a specialized algorithm, named LOC (see Section 3.5.2). In the next section, this new MOO is incorporated into ROPAR to find robust configurations of the network.

To this end, this section describes the problem statement, experimental setup, the case studies, results, analysis and conclusions.

7.1.1 Problem statement

Problem description

Many water quality problems in Water Distribution Networks (WDNs) can be associated to high water residence time in the system, known also as water age. Indeed, high water age affects several physical, biological and chemical aspects, contributing to the accumulation of sediments in pipes, corrosion, undesirable odours and stimulation of chemical reactions (Association 2002; Machell et al. 2009; Machell and Boxall 2014; Masters et al. 2015). Besides, chlorine residual concentration decreases with age, impacting its efficiency as purifier. In distribution systems where chlorine is used as disinfectant, high water age also facilitates the generation of disinfection by-products (DBPs) such as trihalomethanes (THMs), which are carcinogenic (Morris et al. 1992; Weinberg et al. 2002). For these reasons, water age is often taken as a global indicator of water quality in WDNs (Fu et al. 2012; Shokoohi et al. 2017), implying that a well-performing WDNs should keep water age in the network at low values.

The main factors contributing to increase water age are the design of WDNs to cope with future demand based on an estimated growth of the population, possible commercial and industrial developments, and fire protection. For example, in the first years of operation the networks may be oversized for the reduced water demand in comparison to the design demand; the same can happen if the projections of population growth or developments used in the design do not occur as foreseen.

Proper operational interventions, including valve management, can be achieved by means of optimisation procedures. Optimisation has been applied to solve a range of problems related to the design and operation of WDNs and a comprehensive review is reported by Mala-Jetmarova et al. (2017).

Several studies have proposed the optimisation of valves' operations to improve water quality. Prasad and Walters (2006) suggested to minimise water age by finding optimal operational valves status using a single-objective optimisation problem formulation solved with genetic algorithms. More recently, Quintiliani et al. (2017) addressed the same problem using a multi-objective optimisation formulation, in which both the water age and the number of operational interventions are minimised. Abraham et al. (2017) used valve management to maximise the self-cleaning capacity of the network to decrease the risk of discoloration during the peak hours of demand using a single objective optimisation. Other authors have proposed the optimisation of valves' configuration by minimising operational costs. Carpentier and Cohen (1993) optimised the scheduling of valves by the decomposition and coordination of local problems using discrete dynamic programming. Optimal scheduling of valves, among other network elements, has been solved using augmented Lagrangian method (Ulanicki and Kennedy 1994) and using decomposition, using the projected gradient and the complex methods (Cohen et al. 2000a; Cohen et al. 2000b; Cohen et al. 2009). Samora et al. (2015) and Samora et al. (2016) found the optimal location of micro-hydropower turbines with the double purpose of producing electricity and reducing the pressure excess in the network.

Problem formulation

This problem is expressed in its three components: objective functions, decision variables, and constraints.

Objective functions

Two objective functions are considered in the formulation of the optimisation problem. The first objective function (*ObF*1) is to minimise water quality vulnerability in the network, following a similar approach by Di Cristo et al. (2015b), where water quality was assessed based on the concentration of trihalomethanes. In the present study, however, a more general but effective way to represent water quality in a system has been selected. It consists of the minimisation of water age values for each demand node, in an extended period simulation, using the Epanet 2.0 hydraulic solver (Rossman 2010). The estimation of users' exposure to poor water quality can be obtained in different ways. Here, the following three formulations are explored as first objective function (ObF1), one at a time:

1. Maximum water age (*MaWA*), representing the maximum age that occurs during the simulation period across all demand nodes:

$$ObF1 = MaWA = \max WA_{i,t} \ \forall i = 1 \ldots T_n, t = 0 \ldots TST \qquad (48)$$

2. Mean water age (*MeWA*), representing the arithmetic average of the ages at all nodes:

$$ObF1 = MeWA = \frac{1}{T_n * T_{step}} \sum_{i=1}^{T_n} \sum_{t=0}^{TST} WA_{i,t} \qquad (49)$$

3. Demand weighted mean water age (*DeMeWA*), representing the average of the ages calculated assigning a weight to the residence time of water at each node based on the demand requested at each time step:

$$ObF1 = DeMeWA = \frac{\sum_{i=1}^{T_n} \sum_{t=0}^{TST} WA_{i,t} * q_{i,t}}{\sum_{i=1}^{T_n} \sum_{t=0}^{TST} q_{i,t}} \tag{50}$$

where $WA_{i,t}$ is the water age at the i-th node at time step t; T_n is the number of demand nodes of the network, T_{step} represents the number of time steps in which the total simulation time (TST) is divided and $q_{i,t}$ is the demand requested at each node i at any time step t. For the evaluation of these three objective functions, complete mixing at nodes is assumed and dispersion is neglected (Boccelli et al. 1998; Di Cristo and Leopardi 2008). The performance of these three objective functions are compared, in order to compare their effect on water quality in a water distribution system.

The second objective function ($ObF2$) is to minimise the number of valve closures (NoC), defined as the number of existing valves to be closed to reroute the flow in the network. The aim is to limit the interventions in the network to avoid stagnant water and to reduce the investment for the utilities to place new valves.

$$ObF2 = \min NoC \tag{51}$$

Decision variables

Water age is one of the most used indicators of water quality deterioration within a distribution network, which can be modified by the use of control valves to reroute the flows in the system. Hence, the decision variables in the optimisation problem are the statuses of the valves, represented by binary values (1 if the pipe is open and 0 if it is closed). In particular, it has been assumed that each pipe is a potential valve to be operated. It is worth saying that the pipes with the least diameters, used to distribute water to the users, are not considered as decision variables.

Constraints

The operation statuses of the valves need to guarantee the required service standards in terms of quantity, quality and pressure. In terms of water quality, a minimum disinfectant residual concentration has to be maintained at the consumption points, with values that vary according to the relevant regulations, between 0.2mg/L and 0.5mg/L (Organization 2004). The value used follows the Italian regulation, i.e., $[Cl_2]_{i,t}^{min} \geq 0.2 \, mg/l$.

As the EPANET Programmer's Toolkit allows to consider only one parameter at a time for the water quality simulation, a first order kinetic (i.e. Equation (52)) has been used to evaluate chlorine concentration decay, by assuming that the disinfectant is consumed only through a bulk reaction.

$$[Cl_2]_{i,t} = [Cl_2]_{i,t} * e^{-k_b * t} \tag{52}$$

where t is time (h) and k_b is the bulk chlorine decay coefficient, which depends on many factors such as the system hydrodynamics and the environmental characteristics. k_b can be fixed considering literature suggestions or calibrated using measured data. In terms of water quantity, the constraint is that any valve configuration status must guarantee the supply of water to all nodes, i.e., nodes cannot be disconnected.

Finally, the constraint in terms of pressure is that it must be within a fixed range at all nodes and times:

$$P_{min} < P_{i,t} < P_{max} \tag{53}$$

The values used as pressure thresholds are respectively 10 m and 100 m; it means that any valve configuration status that results out of the fixed range will be discarded from the solutions.

Assumptions

To carry out the experiments for this case study, some simplifying assumptions were considered.

Even if in real WDN users are placed along pipes, demands are assumed to be concentrated in nodes. Further investigations will consider demands distributed along pipes as in Farina et al. (2014) and Menapace et al. (2018). For the mean pipe length of the presented networks the corresponding approximation of water age is on the order of less than 1 s.

The pressure-driven approach is not used because the minimum pressure value in the constraint (Equation 53) is fixed in order to guarantee demand-driven functioning.

Leakages are neglected even if they represent a component of demands. Their effect may be analysed in future research.

To verify the existence of disconnected nodes, a procedure implemented in EPANET is used. However, other methods could be also adopted (Creaco et al. 2012).

For water age evaluation complete mixing at nodes is assumed and dispersion is neglected. Although this assumption may be questionable (Machell et al. 2009), its correction requires more complex computations, and for this reason they are still adopted in the majority of simulation tools and applications (Boccelli et al. 1998; Di Cristo and Leopardi 2008; Seyoum and Tanyimboh 2017).

7.1.2 Experimental setup

This section has three parts. In the first part, the main components of the experimental setup and their interconnection is explained. The second part is focused to explain the coupling of the optimiser with the physical modeller. The third part describes the optimisers used.

7.1.2.1 General experimental setup

The proposed experimental setup aims to evaluate different optimisation formulations to be solved by different optimisation algorithms and apply them in two network systems, providing

101

as outputs Pareto fronts and maps to be compared and analysed in order to define the best approach (Figure 33).

The optimisation algorithms used to solve the optimisation problem are the Random Search (RS), the Loop for Optimal valve status Configuration (LOC) and then a combination of each of them with AMGA2. In total, the performance of 4 optimisers is considered. These optimisers will be described in detail the following paragraphs.

The objective functions considered, in addition to the number of closed valves, are the maximum water age (*MaWA*), the average water age (*MeWA*) and the weighted average water age (*DeMeWA*), already introduced in the previous section.

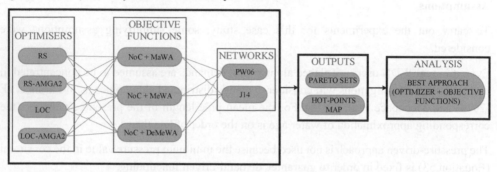

Figure 33. Framework of the experimental setup

Table 11 shows the experiments done for both considered case studies. In particular, each experiment is using a different combination of the optimiser and of the formulations already described for the ObF1 (e.g. P6 is characterized by the use of LOC and *MeWA*).

Table 11. Identification of the optimisation experiments that combine different objective functions and different optimisers

Optimiser	Objective function		
	Maximum water age (*MaWA*)	Mean water age (*MeWA*)	Demand weighted mean water age (*DeMeWA*)
RS	P1	P5	P9
LOC	P2	P6	P10
RS-AMGA2	P3	P7	P11
LOC-AMGA2	P4	P8	P12

7.1.2.2 Model-based optimisation framework

A standard model-based optimisation framework is used. Which consists of two main sections as summarized in Figure 34. Section 1 consist of an executable file compiled in C++, which uses the library of functions of the EPANET Programmer's Toolkit, to set up the valve configurations in the EPANET input file of the network, run the hydraulic and water quality engines, read the model results and calculate the objective functions. Apart from the .inp file of the network, input data include the set of pipes that can be considered as valves and their initial status, which are updated by the optimiser in Section 2.

Section 2 mainly consists on an optimiser algorithm, which reads the values of the objective functions calculated in Section 1, finds the optimal values of the decision variables (i.e., new sets of valves to be closed) and writes their new statuses in a text file. This text file is used to configure a new network with the new pipe statuses in Section 1.

The term 'optimiser' is used to describe any optimisation algorithm. A detailed explanation of the algorithms used are presented in the next section.

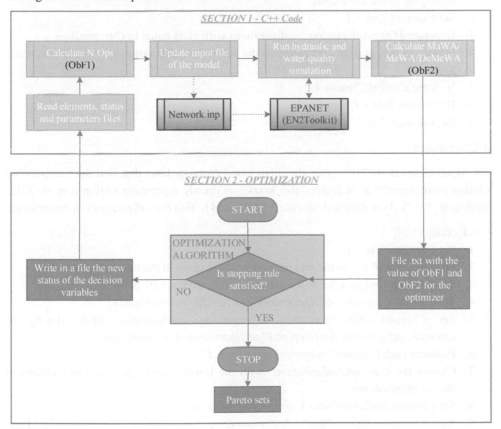

Figure 34. Implementation of the model based optimisation

7.1.2.3 Optimisation algorithms

In this part, the pseudo code of the optimisers Random Search and LOC are explained. Although AMGA2 is other of the optimisers used is not explained in this section because it was already described in Section 2.1. In the last part, the combination of both Random Search and LOC with AMGA2 is detailed.

Random Search

Given a maximum number N of evaluations of the objective function and a maximum number P of pipes to close, RS finds P configurations of the network. In consequence, the first configuration of the network has only one pipe closed (to be named class 1 network), the second configuration has 2 pipes closed (to be named class 2 network), and so on, until the P_{th} configuration, with P pipes closed (to be named class P network), is reached.

Pseudocode:
1. Set $M = N / P$. Where M is the number of network configurations to evaluate all of them belonging to the same class
2. Set *CurrentClass* = 1
3. Generate M network random configurations with class equal to *CurrentClass*
4. Include in the set *BestConfigurations* the network configuration with the lowest water age
5. Increase *CurrentClass* by 1
6. If *CurrentClass* > P then finish
7. Repeat steps 3 to 7

LOC algorithm

LOC (Quintiliani et al. 2019) is an algorithm based on procedures that find the best possible solution incrementally at each step, also known as greedy algorithms (Alfonso et al. 2013; Banik et al. 2017). As in the previous case, LOC is used to find P configurations of the network.

Pseudocode:
1. Set *CurrentClass* = 1
2. Set *RemainingPipes* as the set containing all the pipes in the network.
3. Set *CurrentConfiguration* as the network configuration with all the pipes open.
4. Set C as the cardinality of *RemainingPipes* ($C=|RemainingPipes|$).
5. Set *CurrentConfiguration$_i$* as the resulting configuration after closing in *CurrentConfiguration* the i-th pipe *Pipe$_i$* taken from *RemainingPipes*.
6. Evaluate each *CurrentConfiguration$_i$* for $i = 1..C$.
7. Choose the *CurrentConfiguration$_i$** with the lowest water age which is included in *BestConfigurations*.
8. Set *CurrentConfiguration as CurrentConfiguration$_i$**.
9. Remove the pipe *Pipe$_i$** (corresponding to *CurrentConfiguration$_i$**) from *RemainingPipes*.

10. Increase *CurrentClass* by 1

11. If *CurrentClass* > *P* then finish

12. Repeat step 4 to 12

As it can be inferred from the pseudocode, LOC uses a predetermined, limited number of function evaluations to find a (sub optimal) Pareto front. This number of evaluations is given by the expression:

$$N = \sum_{i=NP-P+1}^{NP} i \qquad (54)$$

where N is the number of function evaluations, NP is the number of pipes of the network and P is the maximum number of pipes to close. $P+1$ becomes the maximum number of solutions in the Pareto front. As it will be explained in the discussion, N will be the basis of comparison among all the optimisation algorithms.

From the previous formula we can derive the running time of the algorithm. Let's consider the worst case that happens when these two conditions are met: first, the algorithm is looking for the maximum number of pipes closed without having any disconnected nodes in the network; second, the WDN can be represented as a fully connected graph. In this case, the worst case, the maximum number of function evaluations are carried out by the algorithm and its order is $O(NP^2)$, which is polynomial.

AMGA2 in combination with other optimisers

Some experiments, not reported here, demonstrated that a randomized search algorithm alone (in this case, AMGA2) was not able to find a satisfactory number of solutions because most of the generated networks had disconnected nodes. To deal with this problem, Prasad and Walters (2006) modified their algorithm to avoid the generation of networks with disconnections. Differently, in this work the search space is reduced to minimise the generation of networks with disconnected nodes by combining AMGA2 with either RS or LOC (named RS-AMGA2 and LOC-AMGA2, respectively). Such approach has led to two important implications for efficiency. First, one of the results, obtained after carrying out the optimisation by either RS or LOC, is the selection of some valves constituting its optimum solution. This selection of valves are the only valves that AMGA2 is going to take into account in the optimisation. By doing this the search space explored by AMGA2 is drastically reduced. Second, this optimum solution by either RS or LOC is taken as the initial population given to AMGA2, improving its efficiency.

Measures to compare the optimisers' performance

In order to measure the improvement of RS and LOC algorithms by combining them with AMGA2, the following *Index of Improvement* (*IoI*) is used:

$$IoI(F_k, F_j) = \frac{1}{|C(F_k, F_j)|} \sum_{C(F_k,F_j)} \frac{f_j^{(1)}}{f_k^{(1)}} \tag{55}$$

where F_k and F_j represent the Pareto fronts of AMGA2 (subscript k) and its counterpart LOC or RS (subscript j), respectively. C is a set containing pairs of solutions, where the first element of the pair is a solution belonging to the Pareto front F_k and the second element of the pair is a solution belonging to the Pareto front F_j. The condition to belong to this set is that the pair must have the same value of $ObF2$. Furthermore, for every pair, $f_k^{(1)}$ is the value of $ObF1$ of the first element of the pair, and $f_j^{(1)}$ is the value of ObF1 of the second element of the pair.

In other words, considering a solution with the same number of operations NoC ($ObF2$), Equation (55) estimates the ratio of the $ObF1$ value of the solution in the counterpart to the $ObF1$ value of the solution with AMGA2. The summation of all these ratios is divided by the number of solutions with the same $ObF2$ to consider a global value representing the efficiency of the procedures, regardless the $ObF1$ formulation used.

Then, the *Weighted Average of the IoI* (*WAIoI*) is evaluated:

$$\tag{56}$$

$$WAIoI(F_k, F_j) = \frac{1}{\sum_{ObF1} |C(F_{k(ObF1)}, F_{j(ObF1)})|} \sum_{ObF1} \left[|C(F_{k(ObF1)}, F_{j(ObF1)})| \right.$$
$$\left. * IoI(F_{k(ObF1)}, F_{j(ObF1)}) \right]$$

where $\sum_{ObF1}$ represents the summation of the sets C for all $ObF1$ formulations.

7.1.3 Case studies

Two case studies with different characteristics are selected to explore the performance of the optimisers and their combinations.

7.1.3.1 Network PW06

This first case study is taken from the work by Prasad and Walters (2006). This is a 47-pipe network that is supplied from a single source, and is formed by 33 demand nodes with elevations that vary between 10 m and 30 m (Figure 35).

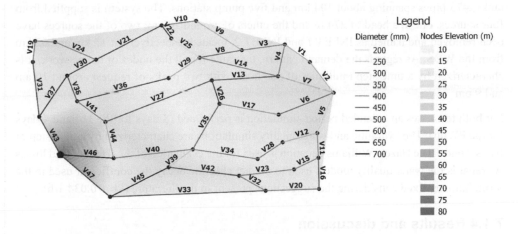

Legend

Diameter (mm)	Nodes Elevation (m)
200	10
250	15
300	20
350	25
400	30
450	35
500	40
600	45
650	50
700	55
	60
	65
	70
	75
	80

Figure 35. Scheme of the PW06 network

The demand pattern assigned to the node for the extended period simulation is constant, in order to compare the results with those obtained by Prasad and Walters (2006).

7.1.3.2 Network J14

The second case study, J14, is a real network from the database developed by the Kentucky Infrastructure Authority (Jolly et al. 2013). The same system has been used as a benchmark test network and updated by different authors (Schal et al. 2016).

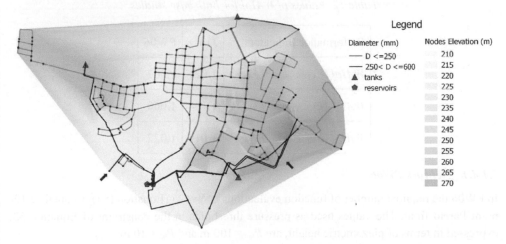

Legend

Diameter (mm)	Nodes Elevation (m)
D <=250	210
250< D <=600	215
▲ tanks	220
⬟ reservoirs	225
	230
	235
	240
	245
	250
	255
	260
	265
	270

Figure 36. Scheme of the J14 network

The model, which can be found in https://zenodo.org/record/437778#.WSvhuOt97IV, has the following characteristics: 316 demand nodes (with elevations between 200 m and 270 m), three tanks, 473 pipes spanning about 104 km and five pump stations. The system is supplied from four sources, one at a head of 274 m and the others of around 200 m; two of the sources have been removed and named as INLET 1 and INLET 2, located respectively at 12 km and 62 km from the WDS. As regards the demand pattern, the same for all the nodes of the network, it is characterized by a time step multiplier of one hour, with two peaks of request around 10 am and 9 pm.

For both test cases an extended period simulation is performed (8 days for the J14 and 4 days for the PW06). The hydraulic and water quality simulations are characterized by a time step of five seconds. The Hazen-Williams equation is used for the evaluation of the friction head losses. As regards the water quality models used to predict chlorine decay the coefficient used in the formulation is fixed considering the suggestions present in the literature (kb = 0.034 1/h).

7.1.4 Results and discussion

The LOC algorithm requires a predefined number of evaluations N (Equation (54)). On the contrary, the other algorithms do not use a predetermined N, which means that their performance directly depends on the required function evaluations. The analysis of the performance is done considering the fixed N of LOC as the baseline.

As more detailed described in the next paragraphs, Figure 37 and Figure 39 show the results of the experiments listed in Table 11 in terms of Pareto fronts for both case-studies, while Table 12 reports the values of the indicator $WAIoI$ (Equation (56)) used to evaluate the performances of the optimisation algorithms.

Table 12. Values of WAIoI for both case studies

Performance Indicator	J14	PW06
$WAIoI$ ($F_{LOC-AMGA2}$, F_{LOC})	1.021	1.007
$WAIoI$ ($F_{RS-AMGA2}$, F_{RS})	1.134	1.060
$WAIoI$ ($F_{RS-AMGA2}$, F_{LOC})	1.010	1.022

7.1.4.1 Network PW06

In PW06 the required number of function evaluations is $N=425$ (Equation (54)) to obtain a 10-point Pareto front. The values used as pressure thresholds in the constraint of Equation 53, expressed in terms of piezometric height, are $P_{max}=100$ m and $P_{min}=10$ m.

For PW06, the solutions reported in terms of Pareto fronts in Figure 37 show that for all considered $ObF1$ formulations, LOC generates a better front than the one from RS. Moreover,

RS and RS-AMGA2 algorithms are able to find a limited number of solutions with respect to LOC and LOC-AMGA2.

AMGA2 barely improves the Pareto front found by LOC. However, its improvement over RS is significant. In fact, the use of AMGA2 in combination with RS allows to reach the same *ObF1* values of RS by operating less valves. Moreover, this combination is also slightly better than LOC and LOC-AMGA2 solutions. This is confirmed by the *WAIoI* values reported in Table 12, which suggest that the addition of AMGA2 produces an improvement of 6.0% and 0.7% with respect to the solutions of RS and LOC, respectively, while the Pareto front of RS-AMGA2 is about 2% better than the one from LOC.

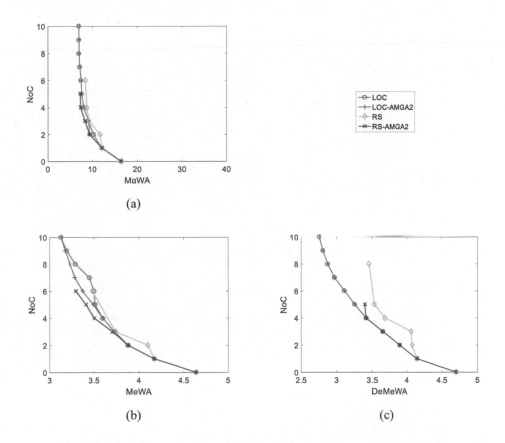

Figure 37. Results for the network PW06 (a) experiments P1 to P4; (b) experiments P5 to P8; (c) experiments P9 to P12

Figure 38 represents for all procedures the heat maps showing the frequency of the valves included in the solutions of the Pareto fronts; a darker dot indicates that the valve is more often considered. RS algorithm (P1-P5-P9) is characterized by the use of a large number of valves in the network, which is not convenient in the operational context. The application of AMGA2 after RS (P3-P7-P11) improves the solutions, focusing only on five or six valves to operate. LOC algorithm has a better behaviour also without the necessity to apply AMGA2 afterwards. Moreover, LOC and LOC-AMGA2 consider almost the same valves, mainly placed on the largest diameters.

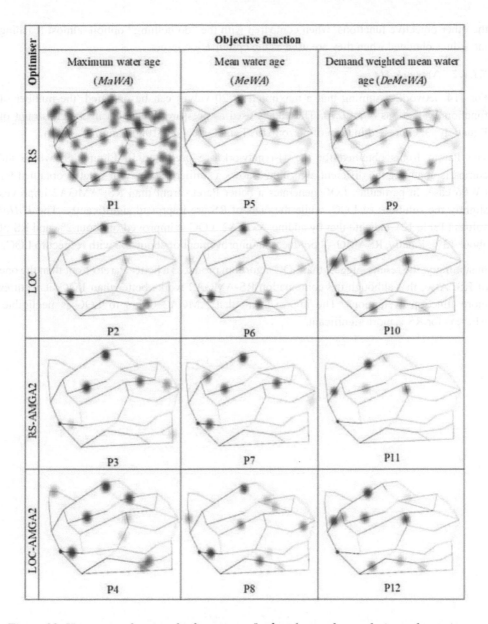

Figure 38. Heat maps showing the frequency of valve closure from solutions of experiments P1 to P12 (the redder, the more frequent)

The performance of each *ObF*1 is also estimated extracting the optimal network configurations and evaluating how well they performed for the remaining *ObF*1 formulations. It is observed that the use of each of the ObF1 formulations implies, on average, a reduction in the values of the other objective functions, when compared with the "do nothing" option, almost reaching the values obtained when they are used as the optimisation target.

7.1.4.2 Network J14

For J14 network, assuming that a maximum of 20 valves can be operated, the number of function evaluations N is 9270. The values used as pressure thresholds in the constraint of Equation 53, expressed in term of piezometric height, are P_{max}=100 m and P_{min}=10 m.

The Pareto fronts obtained for the J14 network are presented in Figure 39, where the comparison among the different algorithms shows a similar tendency of what is obtained for PW06 case. In particular, LOC generates a better Pareto front than RS; AMGA2 improves slightly the solutions of LOC, while the ones of RS are improved significantly. The *WAIoI* values (Table 12), indicate that by adding AMGA2, LOC is improved of about 2% and RS of about 13%. Finally, RSAMGA2 produces an improvement of about 1% with respect to LOC.

In summary, the results suggest that LOC algorithm produces a better Pareto front than the one of RS. Also, that although the combination RS-AMGA2 works better than LOC, it requires more function evaluations. The improvement that AMGA2 offers to LOC is negligible, whereas for RS is more significant.

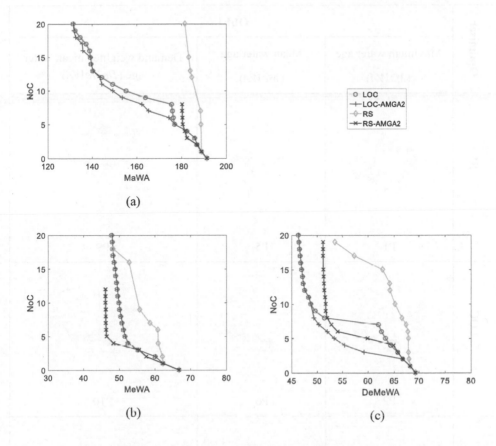

Figure 39. Results for the network J14, experiments P1 to P4 (a); experiments P5 to P8 (b); and experiments P9 to P12(c)

Figure 40 shows the heat maps to provide a spatial indication of where and how frequently the pipes were selected by different procedures (Table 11). As expected, the solutions using the RS algorithm (P1, P5 and P9) do not focus in specific sectors of the network, as the closures are randomly spread over the whole system. Independently from the selected *ObF*1, around 33% of the valves are included in at least one of the solutions, which means that RS requires a large number of valves to be operated.

The solutions obtained with the algorithms RS-AMGA2, LOC and LOC-AMGA2 are characterized by a reduced selection of valves to close, varying from 3% to 4.2% among all the possible decision variables. This confirms again that AMGA2 improves significantly RS. A closer look to the valves selected in each experiment allows to note that RS-AMGA2 individuates different areas respect to LOC and LOC-AMGA2. For the latter algorithms the considered valves are concentrated in specific areas of the network involving mainly the bigger diameters located in the southern part of the system.

113

Optimiser	ObF1		
	Maximum water age (*MaWA*)	Mean water age (*MeWA*)	Demand weighted mean water age (*DeMeWA*)
RS	P1	P5	P9
LOC	P2	P6	P10
RS-AMGA2	P3	P7	P11

Optimiser	ObF1		
	Maximum water age (*MaWA*)	Mean water age (*MeWA*)	Demand weighted mean water age (*DeMeWA*)
LOC-AMGA2			
	P4	P8	P12

Figure 40. Heat maps showing the frequency of valve closure from solutions of experiments P1 to P12 for network J14 (the redder, the more frequent)

About the performance of the *ObF*1 formulations, extracting the optimal network configurations and evaluating how well they performed for the remaining set of *ObF*1 not selected, the results show mixed behaviours. Considering the configuration valves sets obtained using *MaWA* as *ObF*1, it leads to almost no improvements for the other formulations respect to the case of *NoC* = 0. This has serious consequences for the majority of users, because minimising *MaWA* does not imply reduction of the residence time for a large part of the WDN. The solutions obtained with *MeWA* do not modify the values of *MaWA* but improve the ones of *DeMeWA*. This means that the majority of users would have a partial improvement, but not the ones with high water residence time. Similarly, for the solution with *DeMeWA*, *MaWA* remains, on average, near to the zero-closures values regardless the number of closures, while *MeWA* is reduced up to optimal levels. Then, this means that most of the users would have access to water with a reduced age.

7.1.4.3 Performance of the LOC algorithm

In order to evaluate the performance of LOC algorithm, its results are compared with the method proposed by Prasad and Walters (2006) and the Brute-Force Search (BFS) procedure. Those tests have been executed considering the PW06 network and fixing the constraint of 15 meters as the minimum head in the network in accordance with the value used by Prasad and Walters (2006).

The comparison of the results by Prasad and Walters (2006) with those of LOC is shown in Figure 41. For the *MaWA* function, LOC finds several solutions that achieve a similar reduction in water age with less pipe closures. Using the objective function *MeWA* (Figure 41(b)), the

115

LOC solution with 9 closures is as good as the solution of Prasad and Walters (2006) with 11 closures. For *DeMeWA* (Figure 41(c)), LOC with 10 operations marginally dominates the solution by Prasad and Walters (2006). Unfortunately, Prasad and Walters (2006) do not make any reference to the number of evaluations required to get their results so the efficiency of the algorithms cannot be fully compared.

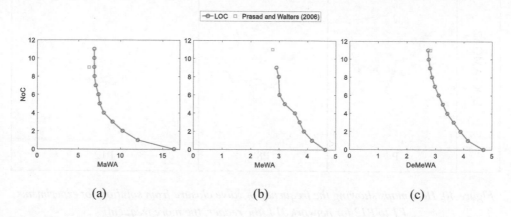

(a)   (b)   (c)

Figure 41. Comparing solutions by Prasad and Walters (2006) with those of LOC using MaWA (a), MeWA (b), and DeMeWA (c)

A further experiment has been designed to prove that the LOC method is suitable to find a close-to-optimal solution. An exhaustive search of all solutions was realized with a Brute-Force Search (BFS) in the smallest network PW06, taking into account *DeMeWA* as ObF1. To reduce the execution time, an array of 28 CPU cores is used to perform the simulations in parallel. Both BFS and LOC were run for 8 pipe closures to achieve the *DeMeWA* maximum reduction.

The solution found by BFS reduced the water age down to 2.8735 h and it was available after 16.6 days of computational effort. Outstandingly, the solution found by LOC reduced the water age down to 2.8736 h, requiring only 3 seconds. This demonstrates the efficiency of the proposed LOC algorithm.

To ensure the reliability of this comparison, the experiment was repeated considering different pipe closures, from 1 to 7. The results are reported in Table 13. In all cases LOC performed as good as BFS, with an advantage of several orders of magnitude in terms of computational time. Unfortunately, it was not feasible to run BFS for NoC of 9, 10 and 11. Indeed, they would take 55 days, 145 days, and 299 days, respectively, because the required number of simulations are 5.44×10^8, 1.44×10^9, and 2.97×10^9, respectively. Moreover, when LOC runs for X closures, the solutions for X-1, X-2,... 2, 1 are immediately available, contrasting with BFS, which requires a separate experiment for each number of closures.

Table 13. Results of the comparative analysis between BSF and LOC, in terms of Water age value, Required number of simulation and Computational time

Number of closures (NoC)	Water age (hr) found		Number of simulations		Computational time (days)	
	BFS	LOC	BSF	LOC	BFS	LOC
1	4.1482	4.1482	4.70E+01	4.70E+01	4.73E-06	4.73E-06
2	3.8869	3.8869	1.07E+03	9.30E+01	1.07E-04	9.36E-06
3	3.6402	3.6402	1.55E+04	1.38E+02	1.56E-03	1.39E-05
4	3.3797	3.4119	1.61E+05	1.82E+02	1.62E-02	1.83E-05
5	3.1795	3.2528	1.28E+06	2.25E+02	1.29E-01	2.27E-05
6	3.0672	3.1072	8.02E+06	2.67E+02	8.07E-01	2.69E-05
7	2.9670	2.9670	4.03E+07	3.08E+02	4.06E+00	3.10E-05
8	2.8735	2.8736	1.65E+08	3.48E+02	1.66E+01	3.50E-05

7.1.4.4 Discussion

From the analysis of the experiments P1 to P12 over the networks PW06 and J14, it is possible to observe the following points:

- LOC produces a better Pareto front than the one produced by RS.
- The improvement of the Pareto front by LOC-AMGA2 with respect to that of LOC is noticeable but not high.
- The improvement of the Pareto front by RS-AMGA2 with respect to that of RS is significant.
- The Pareto front by RS-AMGA2 is slightly better than that by LOC (maximum 2% better according to tests), but at the expense of doing 100% more function evaluations.

All this allows to state that the ideas implemented in the LOC algorithm makes it possible to considerably reduce computational complexity, and if time allows, to achieve certain gains employing the LOC-AMGA2 algorithm. Speed of the employed deterministic optimisation algorithm is also of paramount importance for efficiency of ROPAR, which runs it a large number of times.

7.1.5 Conclusions and Recommendations

The present study compares the performances of 12 multi-objective optimisation procedures to optimise valve management in WDNs for improving water quality, evaluated in terms of water age.

The procedures derive from the combination of four different algorithms (RS, LOC, RS-AMGA2 and LOC-AMGA2) and of three water quality objective function formulations (*MaWA*, *MeWA* and *DeMeWA*). Two distribution networks of varying complexity are considered.

The results show that the proposed LOC algorithm always produces better solutions with respect to RS, obtaining smaller age values with the same number of closures. Moreover, the heat maps show that LOC considers candidate valves concentrated in specific areas of the network, which is an advantage for the operators. Its codification is very simple and it has a good compromise between the quality of the Pareto front and the required number of function evaluations.

The alternatives LOC-AMGA2 and RS-AMGA2 offer only a marginal improvement with respect to the solutions found by LOC, at the expense of having double function evaluations. This implies that, for this particular optimisation problem, LOC algorithm is the most convenient. The heat maps obtained with LOC show also that the operation on the bigger pipes are more efficient for the reduction of the water age. The comparison of LOC with BFS demonstrates that, despite its simplicity, LOC achieves near-to-optimal results with a very small computational effort, which justifies its use in large networks.

7.2 Using ROPAR with the new MOO, two objective functions, twenty four sources of uncertainty

In this section, the problem stated in the previous section is also solved, but with assumption of uncertainty, and the use of robust optimisation. One of the main sources of uncertainty in a WDN is the water demand by the users of the network, and it is considered in this section.

To this end, this section describes the problem statement, experimental setup, the case studies, results, analysis and conclusions.

7.2.1 Problem statement

Problem description

Most of the studies mentioned in Section 7.1.1, neglect that the behaviour of the network can be affected by uncertain parameters or design variables. It is well known that these uncertain inputs can affect the estimation of hydraulic and chemical processes that are mainly model-based (Pasha and Lansey 2005; Idornigie et al. 2010; Di Cristo et al. 2015a). As a consequence, the solutions obtained from a model-based optimisation are also affected by uncertainty in the parameters. Water demand is one of the main recognised sources of uncertainty. Several works have aimed at the minimisation of cost and the maximisation of WDN robustness or resilience taking into account the uncertainty of water demands. Babayan et al. (2004) used the multi-objective optimiser NSGA-II where, in every stage of the evolution, uncertain parameters are used to identify critical nodes and the most significant variables impacting resilience. Kapelan et al. (2005), Kapelan et al. (2006), and Savic (2006) used robust NSGA-II (rNSGA-II) as the multi-objective optimiser, which takes into account uncertainty in the evaluation of the objective functions. To select a robust solution during the evolution of the algorithm, the uncertainty is incorporated carrying out a small number of samples of the uncertain parameters.

The few researches associated to reducing water age via valve management (Prasad and Walters 2006; Quintiliani et al. 2017) use a deterministic approach assuming a 'representative' daily demand pattern without considering uncertainty. This is a limitation because the uncertainty in water demand affects the hydraulic conditions, impacting water age in the network.

Problem formulation

The problem formulation is very similar the one in the previous section. The problem consists of minimising the water age in a network by operating a convenient set of valves, in such a way that the water age stays as close as possible to its minimum value, however now taking into account that water demand is uncertain. It is assumed that every pipe in the network has a potential shut off valve to be operated. The decision variables are the statuses of the valves, represented in principle by binary values (open or closed). Further investigations will consider degrees of valve closures or openings as suggested by Kang and Lansey (2009) and Ostfeld and Salomons (2006).

Objective functions and decision variables

As before, the multi-objective optimisation problem considers two objective functions: the water age, evaluated as the Demand weighted Mean Water Age (*DeMeWA*, Equation (57)) and the Number of valve Closures (*NoC*, Equation (58)):

$$ObF1 = \min_{x} DeMeWA(x, u) = \min_{x} \frac{\sum_{i=1}^{D} \sum_{t=0}^{T} WA_{i,t}(x, u) * q_{i,t}}{\sum_{i=1}^{D} \sum_{t=0}^{T} q_{i,t}} \qquad (57)$$

$$ObF2 = \min_{x} NoC(x, u) \qquad (58)$$

$WA_{i,t}$ is the water age at the *i-th* node at time step *t*; *D* is the number of demand nodes in the network, *T* represents the number of time steps within the Total Simulation Time (*TST*); $q_{i,t}$ is the water demand requested at node *i* at time step *t*. *NoC* represents the number of valves that have been intervened (closed); *x* is the vector of decision variables, containing one of two possible values for each pipe in the network: 1 if the valve is open and 0 if it is closed. The change with the previous formulation is the existence of uncertain vector parameter *u*, containing demand pattern factors (24 positive values, one per hour of the day), which are assumed to be uncertain.

Constraints

Two constraints are needed to guarantee that: 1) any valve configuration status delivers water to all nodes, i.e., nodes cannot be disconnected; 2) the pressure $P_{i,t}$ in each node i at each time t must be within a fixed, acceptable range:

$$P_{min} \leq P_{i,t}, (x, u) \leq P_{max} \qquad \forall t = 0..T \qquad \forall i = 1..D \qquad (59)$$

It is worth noting that the objective functions and the constraints depend on both the decision variables and the demand pattern.

7.2.2 Experimental setup

The proposed experimental setup is summarized in the following sequence of steps executed for every case study:

1. Deterministic optimisation using LOC(following procedure presented in Section 7.1)
2. Robust optimisation using 'ROPAR AD2' and LOC
3. Comparison between deterministic and robust optimisation solutions

7.2.2.1 Design without uncertainty using LOC

To perform the deterministic optimisation, a fixed 24 hours demand pattern is considered in each node. According to the conclusions in Section 7.1.5, LOC alone is enough to perform this deterministic optimisation. Moreover, the efficiency of LOC to find near optimum solutions is very convenient to carry out the robust optimisation described in the next section.

7.2.2.2 Design with uncertainty using ROPAR

For the considered problem, the objective is to obtain a robust configuration of valves, such that minimises water residence time and keeps it near its minimum even under a large range of possible user demands. To carry out the robust optimisation, the ROPAR algorithm is employed. ROPAR is applied to this case in four parts, namely I) Demand pattern sampling from a distribution; II) solving optimisation problem for each sample and generating thus multiple Pareto fronts; III) Analysis of Pareto fronts; IV) Selection of the robust solution.

Part I. Demand pattern sampling

Because demand node variability is related to the behaviour of each user then it is a stochastic process (e.g. Blokker et al. (2009)). In a stochastic representation of this process the values of the uncertain parameters (in this case, a set of hourly pattern multipliers), are generated by Monte Carlo sampling, obtaining thus n different water demand patterns. To generate these uncertain multipliers, the historical mean demand, named $m_{base,k}$, is used as a base. The m stands for multiplier. The sub-index k indicates the hour of the day (i.e., 1 to 24) the multiplier is applied to. To perform the sampling, the daily variability of residential water demand is simulated, considering two different probability distributions: Normal and Log-Normal

(Tricarico et al. 2007). For both probability distributions, the $m_{base,k}$ is the base demand (i.e. the value used for the DO) and the standard deviation is assumed 20% of the base demand, both base demand and its standard deviation obtained from historical demand. In particular, if the base demand is less than 1.5, the Normal distribution is considered; otherwise the Log-Normal distribution is adopted. The uncertain demand is named $m_{uncertain,k}$, which is modelled as an independent random variable obtained from the distributions mentioned above and represented in Equation (60).

$$m_{uncertain,k} \sim \begin{cases} Normal(m_{base,k}, 0.2^2), & if \ m_{base,k} < 1.5 \\ Log-normal(m_{base,k}, 0.2^2), & if \ m_{base,k} \geq 1.5 \end{cases} \qquad (60)$$

For convenience u is defined as the vector $[m_{uncertain,1}, m_{uncertain,2}, ..., m_{uncertain,24}]$. If the networks are in residential areas, only one vector u is necessary to represent the domestic demand of all nodes in the network, and in this case 24 random variables are considered. If the networks have m different demand patterns, then m number of vectors u would need to be generated, and the number of random variables would be $24*m$.

A common assumption from the modelling perspective (e.g., (Blokker et al. 2009; Abraham et al. 2017)) is that the demands from individual households or connections with similar demand type are aggregated and assigned to a node that represents such an area. Here this assumption is also followed, and therefore small pipes connecting the households to the distribution network are not modelled.

Latin Hypercube sampling is employed. The sample consists of n =1,000 demand pattern vectors u. For each pattern, the optimisation problem is solved using LOC, obtaining 1,000 Pareto-quasi optimal sets.

For the process just described, the *sampling confidence* (i.e. *SC*) can be calculated, which is defined as the percentage of the sample space explored. This *SC* is dependent on the *confidence level* of each of the 24 random variables. The formula relating these concepts is:

$$SC(u) = \prod_{k=1}^{24} confidence \ level(m_{uncertain,k}) \qquad (61)$$

Part II) Solving multiple optimisation problems and generation of Pareto fronts

As described above, this is done by running LOC algorithm for each demand pattern sampled at the previous stage. The result is the generation of n Pareto fronts.

Part III) Visual analysis of Pareto fronts

In this step the family of n Pareto fronts obtained in Part I are analysed. The analysis consists of identifying the value of one objective function for which the value of the other objective function yields the lowest possible variance. In this case, the aim is to identify the value of $ObF2$ (i.e. NoC) that yields the lowest variance for $ObF1$ (i.e. $DeMeWA$). To this end, first an initial NoC value is selected; second, all (n) corresponding values of $DeMeWA$ are extracted from the Pareto fronts, and stored in set S. Third, an empirical probability distribution is built with the values in S, obtaining an approximation of the probability density function characterising $DeMeWA$ for that chosen NoC. Finally, the procedure is repeated for different values of NoC, selecting at the end the one that gives the minimum $DeMeWA$ variance.

Part IV) Selection of the robust solution based on the n *Pareto fronts*

This step is executed for the chosen value of $ObF2$ (i.e. NoC) (and can be repeated for several such values). It consists of two main parts. First, check if S contains repeated solutions; if so, then eliminate them in order to obtain a set of unique solutions S_U. Each unique solution might have associated r repeated solutions. This number r can be treated as the 'frequency' of the unique solution, and it is used in the analysis of results.

Second, find a network configuration that works for every possible demand pattern generated, as it is explained next. Take the first value in S_U and retrieve its associated network configuration. For each generated demand pattern $\boldsymbol{u}$, run the retrieved network and obtain its $DeMeWA$ value. If one or more demand patterns generate violations of the pressure constraint, it is said that the network is not reliable and must be discarded; otherwise, this (reliable) network is stored in the set S_R. For each network in S_R, the mean and maximum of $DeMeWA$ (over the n demand patterns) are calculated. Repeat the procedure for the rest of the network configurations associated with the values in S_U. In the end, a set of reliable solutions S_R of cardinality R is obtained. Note that R can be less or equal to the number of Pareto fronts originally generated ($R \leq n$). In addition, two vectors of size R, containing the averages (i.e. A) and the maxima (i.e. M) of $DeMeWA$ are obtained.

Two criteria of robustness are considered: the network that gives the minimum average of $DeMeWA$ (i.e. $ROS1 = \min(A)$), and the minimum maximum of $DeMeWA$ (i.e. $ROS2 = \min(M)$). Note that the solution identified by minimising $ROS1$ can be different from that obtained by minimising $ROS2$. If these solutions are different, it is not possible to generalise that $ROS1$ is always better than $ROS2$ nor vice versa. The selection of one of these two solutions is going to depend on the analysis of the trade-offs of one against the other, a decision that is left to the decision maker.

Furthermore, it is worth mentioning that the minimum maximum $DeMeWA$ is not looking for a solution that complies with a specific threshold. What the method is looking for is a solution that still works well in the worst case of uncertainty realization leading to high $DeMeWA$.

7.2.2.3 Experiments to carry out

The methodology is applied to four different water distribution systems, introduced in the next section, and for each of them *ROS*1 and *ROS*2 are compared. In addition, the networks are solved using deterministic optimisation, and the solution (i.e. *DOS*) is evaluated over the set of *n* demand patterns as the baseline for comparison.

7.2.2.4 Computational complexity

There are two algorithmic parts in the proposed method. The first part is the execution of LOC and the second part is the number of times that LOC is repeated to take into account the uncertainty.

First, the complexity of LOC is analysed. As it can be inferred from its description, LOC uses a limited number of function evaluations to find a Pareto front, understanding as one function evaluation one execution of EPANET. The number of function evaluations is therefore, (Equation (62)).

$$E = \sum_{i=N}^{P} i \qquad (62)$$

where E is the number of function evaluations, P is the number of pipes in the network and N is the number of nodes. The result of this expression is $0.5*[P^2+P-N^2+N]$. Therefore, obtaining a Pareto front using LOC requires polynomial time.

In the second part the LOC algorithm is run for every generated considered sample, and there are 1,000 of them, so the total number of function evaluations is $1000*0.5*[P^2+P-N^2+N]$. Note that the order of this expression is still polynomial.

If there is a possibility to run all these 1,000 executions of LOC in parallel, it would be possible to obtaining the results in the time just a bit higher than the time required for one execution of LOC.

For each of these experiments, 1,000 samples were used. However, in general, the number of samples of *u* depends on two factors, the number of random variables and the size of the network. Reflection on the number of samples is presented in the section discussing the results.

7.2.3 Case studies

Four distribution systems from the literature with different size and topology, representing residential areas, are used to test the methodology. Table 14 contains the main characteristic of these systems. The number that follows the 'Sys' prefix indicates the number of potential valves to be operated in each network. Their schemes are shown in Figure 42. Topology, geometric data, base demand and pattern values are the same as reported in the original papers, and not repeated herein.

123

Table 14. Description of the case studies

	Sys473	Sys365	Sys47	Sys40
Reference	Jolly et al. (2013)	Alfonso et al. (2009), Alfonso (2006)	Prasad and Walters (2006)	Alfonso et al. (2009)
Drawing	Figure 42(a)	Figure 42(b)	Figure 42(c)	Figure 42(d)
Number of pipes	487	366	47	41
Number of potential valves	473	365	47	40
Number of nodes	316	247	33	25
Number of reservoirs	4	3	1	1
Number of tanks	3	0	0	0

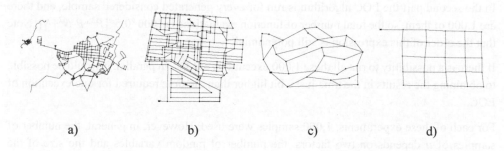

a)   b)   c)   d)

Figure 42. Diagrams of the networks: a) Sys473, b) Sys365, c) Sys47 and d) Sys40

In each considered system, all demand nodes have the same base demand and a one-hour time step pattern. The DO is carried out considering the 'base' demand pattern, while in the RO, the multipliers are modified within a Monte Carlo method, as described in the methodology. For each pattern, hydraulic and quality simulations are performed with a time step equal to five minutes. The *DeMeWA* values (i.e. Equation (57)) have been calculated on a 24 h period (*TST*=24 h). The values used as pressure thresholds in Equation 53 are P_{min}=15 m and P_{max}=100 m, and they guarantee a demand-driven functioning.

7.2.4 Results and discussion

7.2.4.1 Optimisation without uncertainty (LOC approach)

The Pareto fronts obtained with DO for the four networks are shown in Figure 43. It can be observed that DO allows to have, with an adequate number of closures, a reduction of water age of around 50% of the value corresponding to the 'do nothing' option. In evaluating the Pareto front, the 'do-nothing' solution, corresponding to $NoC=0$, is always included to compare how much $DeMeWA$ improves with respect to the original status. The shape of the Pareto front for network Sys473 indicates that the main reduction in $DeMeWA$ is reached with a limited number of closures, while for the others at least one-third of the valves have to be operated.

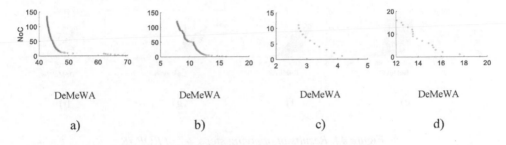

Figure 43. Pareto front of the deterministic optimisation of a) Sys473, b) Sys365, c) Sys47, and d) Sys40

7.2.4.2 Optimisation with uncertainty (ROPAR approach)

Part I and Part II of ROPAR. Figure 44 reports the 1,000 Pareto fronts generated with the RO for each system.

Part III of ROPAR. Although any value of NoC can be selected, two values of NoC have been considered to illustrate the analysis. The first value ($NoC1$) is the number of valve closures for which solutions are available for all Pareto fronts and the minimum value of $DeMeWA$ is reached. These number of closures are 122, 116, 6 and 15 for the networks Sys473, Sys365, Sys47 and Sys40 respectively (see first row of Figure 44).

The second value ($NoC2$) is the 'inflexion point' at which an increment in the number of closures does not represent a significant reduction of $DeMeWA$. These values are 20, 20, 3 and 9 for the networks Sys473, Sys365, Sys47 and Sys40 respectively (see second row of Figure 44).

For the two considered NoC values $NoC1$ and $NoC2$, the PDFs of the 1,000 $DeMeWA$ values are generated, see Figure 44. The solutions corresponding to low NoC values have smaller standard deviations than those with higher value, meaning that they have less variability and therefore they offer more robustness. However, solutions with high NoC values provide a high reduction of $DeMeWA$. This clearly represents a trade-off between having less water age and having less variability. To have more information about this trade-off, it is necessary to carry out the analyses that follow.

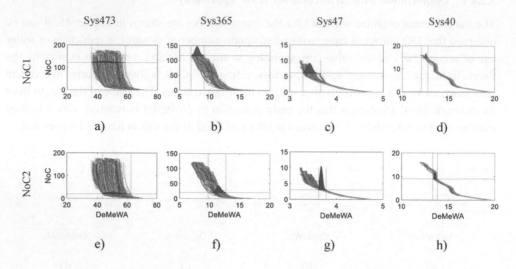

Figure 44. Results of applying steps 4-7 of ROPAR

Part IV of ROPAR. To reduce the computational cost, the analysis related to the repeated solutions is avoided. Therefore only the group of unique solutions for fixed values of *NoC* (*NoC*1 and *NoC*2) is considered (see Table 15). Note that the number of repeated solutions, which is the complement to 1,000 of the unique solutions, decreases with the complexity of the network. The number of repeated solutions is 0% and 16.4% in the Sys473 network for the case *NoC*1 and *NoC*2, respectively, while for the smallest Sys40 the repeated solutions are 98.4%, and 99.3% for *NoC*1 and *NoC*2, respectively.

Figure 45 furnishes information about the reliability and frequency of the unique solutions. The reliability of a solution is the percentage of scenarios where the solution copes, see Section 3.7. To visually present the information more understandably, for each case, the solutions are first ordered by their reliability and then by their frequency. For example considering the chart Sys365, *NoC*2 (Figure 45(f)) every one of its 39 unique solutions have 100% reliability and the 39th solution has a frequency of 358. To simplify the discussion, in the rest of this section a solution is reliable if its reliability is 100%, otherwise it is unreliable. Table 15 indicates that for the *NoC*2 case, for networks Sys365 and Sys40, all unique solutions are reliable. For Sys40, considering *NoC*1 valve closures, only one reliable solution is found. For Sys473 85% of the solutions are not reliable for both *NoC* values. Finally, only one solution is reliable for Sys47 for both *NoC* values. These results suggest that the number of reliable solutions is influenced by the *NoC* value and it cannot be correlated to the size of the network. That is, the number of reliable solutions depends essentially on the functioning of the scheme corresponding to the considered number of closures. The analysis also indicates that the deterministic approach gives solutions that cannot cope with the variability of the demands.

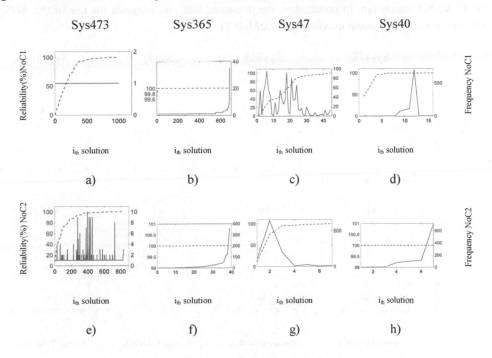

Figure 45. Reliability and frequency of unique solutions

From the reliable solutions, the most robust ones are selected using the criteria $ROS1$ and $ROS2$. Actually because the method is looking for solutions with two criteria of minimum variability, $ROS1$ and $ROS2$, a Pareto front of robust solutions could be produced. Figure 46 shows, for each considered case, the comparison of the solution found by criterion $ROS1$ and $ROS2$ with respect to the rest of the reliable solutions. In three cases represented in Figure 46(c,d,g) only one reliable solution is found. In another three cases (Figure 46(b,f,h)), $ROS1$ and $ROS2$ select the same solution. For the cases reported in (Figure 46(a,e)), different solutions are selected as the most robust from the two criteria. In the case Sys473 and $NoC1$ (Figure 46(a)), a Pareto front with three robust solutions can be seen. These three solutions dominate all the other solutions. The solution in the middle of these three seems to have a good compromise between $ROS1$ and $ROS2$, but ultimately it is left to the decision maker to select of one of these three robust solutions. In the case Sys473 and $NoC2$ (Figure 46(e)), a Pareto front with five robust solutions can be seen.

In summary, the results of the RO show that in six of the eight cases the same solution is identified using both criteria. For the other two cases the $ROS1$ and $ROS2$ solutions have a difference of less than 4% with respect to $DeMeWA$ average, while the difference is about 33%

with respect to *DeMeWA* maximum. This result indicates that the *ROS2* solution is good in terms of *DeMeWA* average, while the *ROS1* one is not as good as the *ROS2* solution considering the *DeMeWA* maximum. In conclusion, the presented analysis suggests the use of the *ROS2* criterion (i.e. the minimum maximum of *DeMeWA*).

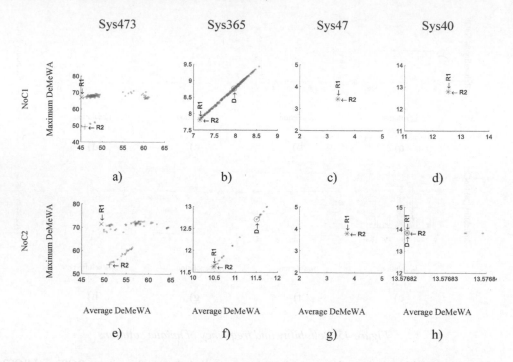

Figure 46. Comparison among ROS1 (R1), ROS2 (R2), DOS (D), and the rest of the reliable solutions for all the cases

7.2.4.3 Discussion

The results for the four networks are analysed next, beginning with the network Sys47. Figure 47 reports the condition with all the valves open and the optimum solutions, deterministic and robust for six pipes closed (i.e. *NoC1*). In the figure, the thickness of each pipe is proportional to its diameter. The case of the network Sys47 is particularly relevant because the deterministic optimisation leads to a solution with 0% reliability (see Table 15). This result shows the importance of performing robust optimisation, because the implementation of a deterministic solution may have undesirable consequences for the system functioning. In contrast, the robust solution has 100% reliability. The water age stays at a minimum value even when the demand varies in a network configuration defined by a robust solution. The maximum *DeMeWA* that this network will have is 3.43 hours and its average *DeMeWA* will be 3.37 hours (see Figure 46(c)). From these values it is clear that even in the worst case, the *DeMeWA* will be just 0.06 hours away from the average, or in other words, the solution stays very close to its minimum

128

value regardless of the variation of the demand. For the case of *NoC*=3, the deterministic solution has 10.7% of reliability (see Table 15). The robust solution has just a difference of 0.06 hours between the average and maximum *DeMeWA*, coinciding with the difference for the case *NoC*=6.

All open   Deterministic   Robust

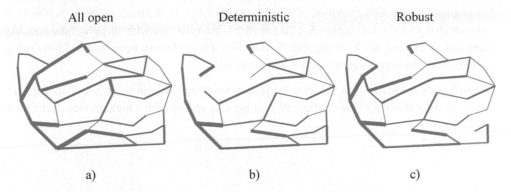

a)   b)   c)

Figure 47. Network Sys47, all pipes open (a), and optimum solutions closing 6 pipes (i.e.
NoC1), deterministic (b) and robust (c)

The results for the network Sys40 illustrate that a deterministic optimisation could find a robust solution for the simplest case of the simplest network (see Figure 46(h)). However there is no way of knowing whether the deterministic solution is indeed robust or not, if not carrying out an analysis considering uncertainty in the demand.

Concerning the network Sys473, there are two important aspects. First point, for both cases analysed (i.e. *NoC*1 and *NoC*2), the solutions found by *ROS*1 and *ROS*2 are different (see Figure 46(a,e)). However just one of the two solutions has to be adopted. In both cases *ROS*2 furnishes a better option because these solutions have the minimum value of the maximum *DeMeWA* and almost the minimum value of the average *DeMeWA*. Second point, from Figure 45(a,e) it is clear that the number of solutions with 100% reliability is very limited. It suggests that for the bigger networks, a solution with 100% reliability may not exist. In this case, it would be desirable or even necessary to increase the number of samples of demand to increase the chances of finding a solution with 100% reliability.

For the network Sys365, from Figure 45(b,f) can be seen that all the solutions have very high reliability (i.e. more than 99.6%). Even the deterministic solutions have 100% reliability (although they are not the most robust). It would be interesting in a future research to explore whether this is due to the architecture of the network.

The *SC* was calculated using Equation (61). The *SC* for the networks Sys473, Sys365, Sys47 and Sys40 are 96.11%, 96.11%, 96.11% and 95.09%, respectively. These percentages can be interpreted as the proportion of probable cases that were tested. This means that the complement of these percentages are the sample spaces not explored, which are 3.89%, 3.89%, 3.89%, and 4.91%, respectively.

Once having the sampling confidence SC (i.e. Equation (61)) and the reliability R (i.e. Equation (6)) of the solution, the total reliability TR (i.e. Equation (30)) of the network can be calculated. For example for the robust solution of the case (Sys473,$NoC1$), the total reliability is 0.9611 x 1.00 which is 96.11%. This means that the network is going to cope with 96.11% of the possible scenarios of demand, while for the other 3.89% ones it is not known because this 3.89% is the proportion of the sample space that was not explored.

Depending on the needs of each network, the number of samples can be increased to augment the SC or the reliability of the configuration or both, to end up with a higher value of the total reliability.

Table 15. Summary of the results for the eight cases considered

	Sys473, NoC=122	Sys473, NoC=20	Sys365, NoC=116	Sys365, NoC=20	Sys47, NoC=6	Sys47, NoC=3	Sys40, NoC=15	Sys40, NoC=9
Unique solutions	1,000	836	683	39	43	7	16	7
Repeated solutions	0	164	317	961	957	993	984	993
Reliable solutions	176	108	675	39	1	1	1	7
No reliable solutions	824	728	8	0	42	6	15	0
Frequency of ROS1	1	1	1	1	12	12	3	713
Frequency of ROS2	1	1	1	1	12	12	3	713
Maximum frequency	1	3	35	358	12	12	3	713
DOS reliability (%)	92.9	94.3	100	100	0	10.7	99.3	100

7.3 CONCLUSIONS

The study presented in this chapter investigates the optimum configuration of valves for keeping the water age in a distribution network to reliable levels to avoid water quality degradation. Four different networks of varying sizes and complexities are used as case studies. The multi-objective optimisation problem is formulated, aimed at minimising the water age (*ObF*1) and the number valve closures (*ObF*2), and solved using the LOC algorithm. The problem is first solved in deterministic setting, and then with the explicit account for uncertainty in demands, using ROPAR approach. For identifying the most robust solution, two different criteria, based on the minimisation of the average and maximum *ObF*1 value, named *ROS*1 and *ROS*2, respectively, are used.

The analysis shows that in many cases the solution(s) found by DO may not satisfy the constraints necessary for an adequate hydraulic functioning of the network if demand values are varied, and implementation of a deterministic solution may lead to the system malfunctioning. It is demonstrated that RO leads to different solutions which are more appropriate to implement in case of uncertainty in demand. Moreover, the RO results indicate that the solutions selected by the criterion based on the minimisation of the maximum $ObF1$ are consistently more robust than those by the other criterion. The analysis also shows that the number of reliable solutions is influenced by the number of closures and is not correlated to the size of the network.

Recommendations for future research are related to the following two points. First, to take into account new designs in valves that can be operated not only in binary states (i.e. closed or opened) but also in intermediate degrees of those two states. Second, to consider that the network has different kinds of users and/or different consumption through the year impacting with this the demand patterns to take into account.

8

CONCLUSIONS AND RECOMMENDATIONS

8.1 SUMMARY

Optimising the solution of water related problems without explicitly taking into account uncertainty of model inputs or model parameters could lead to solutions that underperform under real situations. This thesis focused on three knowledge gaps existing in current optimisation of water related systems, namely the lack of explicit consideration of uncertainty in the existing model-based optimisation algorithms used to solve water-related problems, the lack of uncertainty estimates of the resulting optimal solutions and the need to develop adequate algorithms to solve the problems of robust optimisation. This research addressed these issues focusing on robust optimisation of multiple objectives, to find solutions that are in a certain sense robust, i.e. remain near their optimal performance, across a range of the associated uncertainties.

In this thesis, the Robust Optimisation and Probabilistic Analysis of Robustness (ROPAR) algorithm was developed, implemented and tested. ROPAR is an approach that uses standard robustness criteria that are widely accepted in literature. This standardization helps to increase its usage in different types of problems concerning multi-objective robust optimisation of multiple objectives. Its main feature is the ability to propagate uncertainty of random parameters to the solutions. This allows for having more information about the performance of the solutions under different conditions. Feasibility of ROPAR to find robust solutions in different kind of problems was tested on a number of diverse case studies with different numbers of objective functions, decision variables, uncertain parameters and complexities.

Results show that ROPAR is general enough and could be applied not only to other kind of problems. Additionally, by design, ROPAR algorithm can be straightforwardly distributed to an array of computer to minimise the time to obtain robust solutions.

8.2 CONCLUSIONS

The conclusions are arranged as the answers to the research questions, as presented in Section 1.3.

1. *What algorithm of optimisation of multiple objectives could be used as a 'deterministic optimisation engine' in the robust optimisation framework?*

For the storm drainage systems, NSGAII (Deb et al. 2002) and AMGA2 (Tiwari et al. 2011) were used. NSGAII was selected because it is still the most widely used optimiser of multiple objectives. AMGA2 was selected because it is one of the most efficient algorithms, according to the tests that were carried out although not reported in this thesis. Besides, AMGA2 is open source and readily available online and it does not have any restriction with respect to its use or modification.

For water distribution systems, in particular for the problem of the minimisation of both water age and the number of pipes closed, the straightforward application of any evolutionary algorithm, including AMGA2, was not useful because of the highly constrained nature of the problem most solutions generated during the process of optimisation were invalid (they had disconnected nodes). Looking for an alternative to a genetic algorithm, the LOC algorithm (i.e. Loop for Optimal valve status Configuration, Quintiliani et al. (2019)) was developed and tested in this thesis. LOC is a greedy algorithm capable of finding almost optimal network configurations to reduce water age in the network. Furthermore, LOC is very efficient in terms of the number of simulations carried out to find this optimum network configuration. This efficiency is very convenient given the number of times that LOC has to be executed to carry out the ROPAR analysis.

ROPAR, by design, allows for using any kind of Multi-Objective Optimisation (i.e. MOO) algorithm. In this study two kinds of MOO were used, namely evolutionary algorithms (NSGAII and AMGA2) and the specialized greedy algorithm (LOC).

From the experience obtained during employing ROPAR, it is recommended to search for a MOO suitable to the particular kind of the problem to be solved. If such a MOO cannot be found, then a general MOO can be used, e.g. the ones used in this thesis, the reportedly efficient Borg (Hadka and Reed 2013), etc.

   2. *How to estimate robustness of optimal solutions given uncertainty of inputs or parameters?*

ROPAR indeed allows for analysing the impact of the propagated uncertainty in parameters on robustness of the identified solutions.

The visual analysis allows for identifying the range of solutions that could be less sensitive to uncertainty. This visual analysis consists of inspection of the Probability Density Functions (i.e. PDFs) of the solutions at different objective function levels. The wider the PDF, the more sensitive are the solutions. However, at this point it is neither possible to determine robustness of the solutions, nor to identify which of them are the most robust. This analysis is carried out in the subsequent numerical analysis, where the robustness of the optimal solutions is assessed using four criteria that relate to uncertainty. Although these four criteria of robustness were used because they are recognized and recommended in the literature (Beyer and Sendhoff 2007), the ROPAR framework also allows to incorporate additional robustness criteria.

Once the value of these four robustness criteria is calculated for each of the optimal solutions, the selection of the most robust solutions (and in consequence, those least sensitive to uncertainty), can be carried out.

   3. *What considerations must be taken into account by a generic framework to find robust solutions?*

The notions of robustness adopted in particular domain areas could be quite different, and hence the criteria of robustness used in this thesis. Additionally, ROPAR allows for modification of these criteria.

This framework could be used to solve various problems where uncertainty is defined probabilistically. But the problem should not be based on very complex (long-running) models, otherwise the running time could be unbearable, because this is a computationally intensive framework. Robust optimisation in case of very complex problems would not be accurate, because inevitably the sampling space would be limited.

4. How can definition of robustness be extended to problems with multiple objectives?

In most of the cases found in the literature, the robust optimal solutions comply with one criterion of robustness. The ROPAR framework allows for finding solutions complying with four criteria of robustness (Beyer and Sendhoff 2007). Additionally, ROPAR is extensible to include more criteria of robustness.

This framework is also incorporating additional terms related with robustness such as reliability (Loucks and Van Beek 2017). This concept is related to the fulfilment of the problem constraints and has been also used in some problems of water distribution systems (Kapelan et al. 2005; Savic 2005; Kapelan et al. 2006; Savic 2006).

Additionally, the confidence in sampling is also taken into account in this work to determine the total reliability of the solution. There is no doubt that this total reliability of the solution is also impacting its robustness.

5. How does the proposed algorithm compare with the existing approaches to robust optimisation, and what is its computational complexity?

Because four metrics of robustness are used, the set of robust solutions would form a Pareto front in four dimensions. To visually analyse such Pareto front is challenging because maximum three dimensions can straightforwardly be represented in a graphic. In order to ease this visualization, a method of normalized aggregation of these four robustness metrics is proposed and used.

Using this method to evaluate the robustness, the solutions generated by the Optimisation by Smoothing the Objective Function (i.e. OSOF) were compared with the solutions by ROPAR. According to the comparisons carried out in this work, the solutions found by ROPAR are at least as good as the solutions found by OSOF.

It is however acknowledged that it is necessary not only to test the framework with more complex cases, but also to compare the framework with other approaches to robust optimisation.

6. Can the proposed framework be used for problems of different contexts and settings?

This framework was tested in different kinds of case studies: first, benchmark function; second, storm drainage systems; third, water distribution systems.

ROPAR was initially tested with a benchmark function. This benchmark function, ZDT1 (Zitzler et al. 2000), is traditionally used to test multi-objective optimisation algorithms. Here, the original formulation was modified to introduce uncertainty. This was useful to see the propagation of uncertainty to optimal solutions.

136

Another type of case studies was storm drainage systems. The analysis began using a simple case, useful to make clear the use of the framework. Next, two complex cases were robustly optimised by using two algorithms of robust optimisation and their solutions were compared with respect to robustness. Finally, a more complex drainage system, in conjunction with three objective functions and three uncertainty sources, was useful to test the method with problems of many objectives.

The other kind of case studies was water distribution systems. This type of case brought challenges not seen in the previous cases. First, in the previous cases a genetic algorithm was used as the optimisation algorithm, however, the same algorithm was found not to be useful, and the new optimisation method was developed. This experience confirmed the idea that this framework allows for the use of any kind of multi-objective optimisation algorithm. Second, in the previous cases no constraints were considered, and this case allowed to test the framework on problems with constraints. Third, the number of uncertainty sources, which was small (only up to three in the first cases), was increased to 24, allowing to test the framework for multiple sources of uncertainty.

    7.   What are the ways to ensure reasonable efficiency in obtaining robust solutions?

Most of the existing algorithms to find robust solutions to problems with multiple objectives are very CPU intensive. When these algorithms are executed in one single computer, the process time can be prohibitively long. To decrease this execution time, it is possible to distribute the load among several computers. However, the parallelization of these algorithms is not straightforward.

One of the algorithms that is more efficient is rNSGAII (Kapelan et al. 2005; Kapelan et al. 2006). This algorithm saves time by selecting strategically the samples to use during the robust optimisation process.

ROPAR solves the optimisation problem multiple times, and so by design it is computationally intensive. However, it can be straightforwardly distributed among several computers what allows for decreasing of the execution time.

8.3 LIMITATION OF THIS STUDY

The new framework to find robust solutions was tested on different kind of problems and different kind of settings. However, it would be necessary to consider problems with more diversity and more complexity to verify whether this framework works well in those contexts too.

Regarding the storm drainage case studies, all of them have circular pipes. Additionally, the Colombo case study also has canals. During the optimisations carried out in this thesis, both pipes and canals were considered custom made, meaning that the specific pipe diameter or canal width could be built to the specification determined by the optimisation. This assumption is generally true for canals, however it generally does not hold for pipes. Therefore, in case of applying the methods mentioned in this thesis, this is a point to be taken into account.

For the cases when there is a need to use the available commercial pipes, the design process has to consider these pipe diameters into the optimisation. Because this modification should be carried out only in the optimisation algorithm, the ROPAR algorithm does not require a change; since ROPAR can work with any multi-objective optimisation algorithm.

Another point to consider in the storm drainage case studies is related to the uncertainties taken into account. They were three, namely the rainfall, pipe age and the evolution of imperviousness in the basin. This is of course a simplification since in reality other variables have to be considered. For example, pipe slopes, which have an impact on the flow velocity leading to erosion problems if they are too high or sedimentation if they are too low, excavation costs, etc.

8.4 RECOMMENDATIONS

It should be noted that the capability of ROPAR comes at a cost: since it is based on Monte Carlo framework, and requires solving MOO problem multiple times, it is computationally expensive, so the use of a surrogate model of hydrodynamic models would be advisable to expedite its execution. This surrogate model could be a neural network or any other data-driven model suitable to represent the underlying processes. Using this approach would reduce significantly the execution time.

Although this study proposes a method to calculate the size of the sample, there is no doubt that other methods could also be applied. This is important because the quality of the sample influences trust one can have on the solution. On the other hand, using a small but representative sample could help reducing the computational time to find near-to-optimal solutions. This efficient robust solution finder could be useful in situations where there is pressure to promptly find a solution or in situations where it is very expensive to execute the simulation model calculating the objective functions.

A particular implementation of OSOF, such as rNSGA-II (Kapelan et al. 2005; Kapelan et al. 2006), has the possibility to find close-to-optimal solutions by selectively using a reduced number of samples through the optimisation process. It would be worth exploring the modification of ROPAR to find "close-to-optimal" solutions either by reducing the number of samples or by modifying the framework itself.

The ROPAR framework uses four robustness metrics, however it allows to include other robustness metrics as well, e.g. the eleven ones considered by McPhail et al. (2018). Four robustness metrics out of those eleven are used in this thesis, and it would be worth exploring the remaining seven metrics in the ROPAR framework.

The ROPAR framework could be extended by including sensitivity analysis. Herman et al. (2015) developed a taxonomy of current methods for robust decision making. It has four stages, and the fourth stage is related to sensitivity analysis and could be used as the basis for adding sensitivity analysis to the ROPAR framework.

Future research efforts will be also aimed at testing ROPAR and its potential extensions to deal with multiple sources of uncertainty, tests on more complex cases studies, and considering more objective functions.

LIST OF ACRONYMS

AFV — Average Flood Volume

AMALGAM — A Multi ALgorithm Genetically Adaptive Method

AMGA2 — Archive-based Micro Genetic Algorithm

BMP — Best Management Practices

C-NSGA-II — Clustered Non-dominated Sorting Genetic Algorithm-II

DBPs — Disinfection By-Products

CDF — Cumulative Distribution Function

CPU — Central Processing Unit

CSA — Cross Section Area

DeMeWA — Demand weighted Mean Water Age

DET — Deterministic

DO — Deterministic Optimisation

DOS — Deterministic-Optimal Solution

EA — Evolutionary Algorithm

EPA — United States Environmental Protection Agency's

EPANET — Water distribution system modelling software package developed by EPA

Epsilon-MOEA — Epsilon Multiple Objective Evolutionary Algorithm

Epsilon NSGA-II — Epsilon Non-dominated Sorting Genetic Algorithm-II

ES — Evolutionary Strategy

GDE3 — Generalized Differential Evolution 3

IBEA — Indicator-Based Evolutionary Algorithm

IDF — intensity-duration-frequency equation

ii — index of improvement

J14 — WDN in (Jolly et al. 2013)

LOC — Loop for Optimal valve status Configuration

MaWA — Maximum water age

MeWA — Mean water age

MFV — Maximum Flooding Volume

MLW — Millions of Liters of Water

MMU — Millions of Monetary Units

MOEA — Multiple Objective Evolutionary Algorithm

MOEA/D — MOEA based on Decomposition

MOO — Multi-Objective Optimisation

MPI — Message Passing Interface

NoC — Number of Closures

NP — Number of Pipes

NRI — NormRobustnessIndicator

NSGA-II — Non-dominated Sorting Genetic Algorithm-II

OF — Objective Function

OMOPSO — Optimal Multi-objective Optimisation PSO

OSOF — Optimisation by smoothing the objective function

PC — Personal Computer

PDF — Probability Density Function

PSO — Particle Swarm Optimisation

PW06 — WDN in Prasad and Walters (2006)

RI — RobustnessIndicator

RMOO – Robust Multi-Objective Optimisation

RO – Robust Optimisation

ROPAR — Robust Optimisation and Probabilistic Analysis of Robustness

ROPAR A —ROPAR version to propagate uncertainty to the solutions

ROPAR AD2 — ROPAR version to find most robust solutions by using 2 robustness criteria

ROPAR AD4 — ROPAR version to find most robust solutions by using 4 robustness criteria

ROS — Robust-Optimal Solution

rNSGAII — robust NSGAII

ROS — Robust Optimum Solution

RS — Random Search

SC — Sampling Confidence

SWMM — Storm Water Management Model

THMs — TriHaloMethanes

TMU — Thousands of Monetary Units

TR — Total Reliability

TST — Total Simulation Time

US — Uncertainty Source

USD — US Dollars

WA — Water Age

waii — weighted average of the index of improvement

WDN — Water Distribution Network

WDS — Water Distribution System

WNI — Water Not Infiltrated

WRP — Water Related Problem

ZDT1 — Benchmark function with this name

LIST OF TABLES

LIST OF FIGURES

ABOUT THE AUTHOR

Oscar Osvaldo Marquez Calvo was born in Oaxaca, Mexico. He received his B.Sc. in Computer Systems Engineering in 1990 from the Monterrey Institute of Technology and Higher Education (ITESM is its acronym in Spanish). In 1992 he obtained his MSc in Computer Systems from the ITESM. In 1994 he received his MSc in Automation from the ITESM. After this he worked in the area of information technology supporting and developing software for several private companies and for his own business as well.

In November of 2011 he joined the 'Centro del Agua' which is a water research centre in Monterrey, Mexico. There he developed several computer systems to assist the research of the centre. In November 2012, he started working on his PhD research on robust optimisation at IHE Delft Institute for Water Education, in the Chair group of Hydroinformatics. Part of his time during the first year of the PhD he also contributed to the activities of the European project KULTURisk: he participated in the development of a Web-based Knowledge Base Platform supporting the process of raising flood risk awareness.

JOURNAL PUBLICATIONS

Marquez-Calvo O. O. & Solomatine D. P. 2019 Approach to robust multi-objective optimization and probabilistic analysis: the ROPAR algorithm. Journal of Hydroinformatics 21, 427-440 doi:10.2166/hydro.2019.095.

Marquez-Calvo O. O., Quintiliani C., Alfonso L., Di Cristo C., Leopardi A., Solomatine D. P. & de Marinis G. 2018 Robust optimization of valve management to improve water quality in WDNs under demand uncertainty. Urban Water Journal 15, 943–952 doi:10.1080/1573062X.2019.1595673.

Quintiliani C., Marquez-Calvo O. O., Alfonso L., Di Cristo C., Leopardi A., Solomatine D. P. & de Marinis G. 2019 Multi-objective valve management optimization formulations for water quality enhancement in WDNs. Journal of Water Resources Planning and Management 145, 04019061 doi:10.1061/(ASCE)WR.1943-5452.0001133.

CONFERENCE PROCEEDINGS

Marquez-Calvo, O. O., & Solomatine, D. P. (2015). Towards robust optimal design of storm water systems. Paper presented at the General Assembly 2015 of the European Geosciences Union, Vienna, Austria.

Marquez-Calvo, O. O., & Solomatine, D. P. (2016). Experiments with ROPAR, an approach for probabilistic analysis of the optimal solutions' robustness. Paper presented at the General Assembly 2016 of the European Geosciences Union, Vienna, Austria.

Marquez-Calvo, O. O., & Solomatine, D. P. (2019). Comparative analysis of ROPAR, a method to find robust optimum solutions to problems with multiple objectives. Paper presented at the General Assembly 2019 of the European Geosciences Union, Vienna, Austria.

REFERENCES

Abraham E., Blokker M. & Stoianov I. 2017 Decreasing the Discoloration Risk of Drinking Water Distribution Systems through Optimized Topological Changes and Optimal Flow Velocity Control. Journal of Water Resources Planning and Management 144, 04017093

Akan A. O. & Houghtalen R. J. 2003 Urban hydrology, hydraulics, and stormwater quality: engineering applications and computer modeling. John Wiley & Sons, Hoboken, NJ

Alfieri L., Laio F. & Claps P. 2008 A simulation experiment for optimal design hyetograph selection. Hydrological Processes 22, 813-820

Alfonso L. 2006 Use of hydroinformatics technologies for real time water quality management and operation of distribution networks. Case study of Villavicencio, Colombia. Master's thesis, UNESCO-IHE Institute for Water Education, Delft, The Netherlands.

Alfonso L., He L., Lobbrecht A. & Price R. 2013 Information theory applied to evaluate the discharge monitoring network of the Magdalena River. Journal of Hydroinformatics 15, 211-228 doi:10.2166/hydro.2012.066.

Alfonso L., Jonoski A. & Solomatine D. 2009 Multiobjective optimization of operational responses for contaminant flushing in water distribution networks. Journal of Water Resources Planning and Management 136, 48-58 doi:10.1061/(ASCE)0733-9496(2010)136:1(48).

Andino-Santizo O. F. 2012 Development and application of an optimization tool for urban drainage network design under uncertainty. Master's thesis, UNESCO-IHE Institute for Water Education, Delft, The Netherlands.

Arachchige-Don S. R. 2015 Urban Drainage pipe criticality analysis considering hydraulic performance and maintenance cost. Master's thesis, UNESCO-IHE Institute for Water Education, Delft, Netherlands.

Arnbjerg-Nielsen K., Willems P., Olsson J., Beecham S., Pathirana A., Bülow Gregersen I., Madsen H. & Nguyen V.-T.-V. 2013 Impacts of climate change on rainfall extremes and urban drainage systems: a review. Water Science and Technology 68, 16-28 doi:10.2166/wst.2013.251.

Ashley R. M., Balmforth D. J., Saul A. J. & Blanskby J. 2005 Flooding in the future–predicting climate change, risks and responses in urban areas. Water Science and Technology 52, 265-273

Association A. W. W. 2002 Effects of Water Age on Distribution System Water Quality. American Water Works Association: Denver, CO, USA, 19

Babayan A. V., Savic D. A. & Walters G. A. 2004 Multiobjective optimization of water distribution system design under uncertain demand and pipe roughness. Water Resources Planning and Management 130, 467-476

Banik B. K., Alfonso L., Di Cristo C., Leopardi A. & Mynett A. 2017 Evaluation of Different Formulations to Optimally Locate Sensors in Sewer Systems. Journal of Water Resources Planning and Management 143, 04017026 doi:10.1061/(ASCE)WR.1943-5452.0000778.

Barreto W., Vojinovic Z., Price R. & Solomatine D. 2009 Multiobjective evolutionary approach to rehabilitation of urban drainage systems. Journal of water resources planning and management 136, 547-554

Basdekas L. 2014 Is Multiobjective Optimization Ready for Water Resources Practicioners? Utility's Drought Policy Investigation. J Water Resour Plann Manage 140, 275-276

Beh E. H., Maier H. R. & Dandy G. C. 2015 Scenario driven optimal sequencing under deep uncertainty. Environmental Modelling & Software 68, 181-195

Ben-Tal A. & Nemirovski A. 2002 Robust optimization – methodology and applications. Mathematical Programming, 453-480

Beyer H.-G. & Sendhoff B. 2007 Robust optimization–a comprehensive survey. Computer methods in applied mechanics and engineering 196, 3190-3218

Blokker E., Vreeburg J. & Van Dijk J. 2009 Simulating residential water demand with a stochastic end-use model. Journal of Water Resources Planning and Management 136, 19-26 doi:10.1061/(ASCE)WR.1943-5452.0000002.

Boccelli D. L., Tryby M. E., Uber J. G., Rossman L. A., Zierolf M. L. & Polycarpou M. M. 1998 Optimal scheduling of booster disinfection in water distribution systems. Journal of Water Resources Planning and Management 124, 99-111 doi:10.1061/(ASCE)0733-9496(1998)124:2(99).

Butler D. & Davies J. W. 2004 Urban drainage. Second edn. Spon Press, London, Great Britain

Carpentier P. & Cohen G. 1993 Applied mathematics in water supply network management. Automatica 29, 1215-1250 doi:10.1016/0005-1098(93)90048-X.

Cheng S., Zhou J. & Li M. 2015 A New Hybrid Algorithm for Multi-Objective Robust Optimization With Interval Uncertainty. Journal of Mechanical Design 137, 021401-021401/021401-021409

Chiong R., Weise T. & Michalewicz Z. 2012 Variants of evolutionary algorithms for real-world applications. Springer,

Cochran W. G. 1977 Sampling techniques. Wiley series in probability and mathematical statistics, 3rd edn. John Wiley and Sons, Inc., New York

Cohen D., Shamir U. & Sinai G. 2000a Optimal operation of multi-quality water supply systems-I: Introduction and the QC model. Engineering Optimization+ A35 32, 549-584 doi:10.1080/03052150008941313.

Cohen D., Shamir U. & Sinai G. 2000b Optimal operation of multi-quality water supply systems-II: The QH model. Engineering Optimization+ A35 32, 687-719 doi:10.1080/03052150008941318.

Cohen D., Shamir U. & Sinai G. 2009 Optimisation of complex water supply systems with water quality, hydraulic and treatment plant aspects. Civil Engineering and Environmental Systems 26, 295-321 doi:10.1080/10286600802288168.

Congedo P. M., Witteveen J. A. S. & Iaccarino G. 2011 Simplex-simplex approach for robust design optimization. In: Evolutionary and Deterministic Methods for Design, Optimization and Control (Eurogen 2011), Capua, Italia. Springer,

Cozzolino L., Cimorelli L., Covelli C., Mucherino C. & Pianese D. 2015 An innovative approach for drainage network sizing. Water 7, 546-567

Creaco E., Franchini M. & Alvisi S. 2012 Evaluating water demand shortfalls in segment analysis. Water resources management 26, 2301-2321

Darwin C. 1859 On the Origin of Species by Means of Natural Selection Or the Preservation of Favoured Races in the Struggle for Life. H. Milford; Oxford University Press,

Deb K. & Gupta H. 2006 Introducing Robustness in Multi-Objective Optimization. Evolutionary Computation 14, 463-494 doi:10.1162/evco.2006.14.4.463.

Deb K., Mohan M. & Mishra S. 2005 A fast multi-objective Evolutionary Algorithm for Finding Well-Spread Pareto-Optimal Solutions. Evolutionary Computation 13, 501-526

Deb K., Pratap A., Agarwal S. & Meyarivan T. 2002 A Fast and Elitist Multiobjective Genetic Algorithm: NSGA-II. IEEE Transactions on Evolutionary Computation 6, 182-197

Di Cristo C. & Leopardi A. 2008 Pollution source identification of accidental contamination in water distribution networks. Journal of Water Resources Planning and Management 134, 197-202 doi:10.1061/(ASCE)0733-9496(2008)134:2(197).

Di Cristo C., Leopardi A. & de Marinis G. 2015a Assessing measurement uncertainty on trihalomethanes prediction through kinetic models in water supply systems. Journal of Water Supply: Research and Technology-Aqua 64, 516-528 doi:10.2166/aqua.2014.036.

Di Cristo C., Leopardi A., Quintiliani C. & De Marinis G. 2015b Drinking water vulnerability assessment after disinfection through chlorine. Procedia Engineering 119, 389-397

Erfani T. & Utyuzhnikov S. V. 2012 Control of robust design in multiobjective optimization under uncertainties. Structural and Multidisciplinary Optimization 45, 247–256 doi:10.1007/s00158-011-0693-0.

Farina G., Creaco E. & Franchini M. 2014 Using EPANET for modelling water distribution systems with users along the pipes. Civil Engineering and Environmental Systems 31, 36-50

Fieldsend J. E. & Everson R. M. 2005 Multi-objective optimisation in the presence of uncertainty. In: IEEE Congress on Evolutionary Computation, CEC 2005, Edinburgh, Scotland. pp 243-250

Fortunato A., Oliveri E. & Mazzola M. 2014 Selection of the optimal design rainfall return period of urban drainage systems. Procedia Engineering 89, 742-749

Fu G., Kapelan Z., Kasprzyk J. R. & Reed P. 2012 Optimal design of water distribution systems using many-objective visual analytics. Journal of Water Resources Planning and Management 139, 624-633 doi:10.1061/(ASCE)WR.1943-5452.0000311.

Galindo-Calderon R., Cano C., Sanchez A., Vojinovic Z. & Brdjanovic D. 2015 Selecting optimal sustainable drainage design for urban runoff reduction. In: The Hague, the Netherlands. E-proceedings of the 36th IAHR World Congress,

Gaspar-Cunha A. & Covas J. A. 2008 Robustness in multi-objective optimization using evolutionary algorithms. Computational Optimization and Applications 39, 75-96 doi:10.1007/s10589-007-9053-9.

Gunawan S. & Azarm S. 2005 Multi-objective robust optimization using a sensitivity region concept. Structural and Multidisciplinary Optimization 29, 50-60 doi:10.1007/s00158-004-0450-8.

Guo Y., Walters G. & Savic D. 2008 Optimal design of storm sewer networks: Past, present and future. In: 11th International Conference on Urban Drainage, Scotland, UK.

Hadka D. & Reed P. 2013 Borg: An Auto-AdaptiveMany-Objective Evolutionary Computing Framework. Evolutionary Computation 21, 231-259

Haestad & Durrans S. R. 2007 Stormwater Conveyance Modeling and Design. First edn. Bentley Institute Press, Exton, Pennsylvania, USA

Herman J. D., Reed P. M., Zeff H. B. & Characklis G. W. 2015 How should robustness be defined for water systems planning under change? Journal of Water Resources Planning and Management 141, 04015012

Huff F. A. 1990 Time distributions of heavy rainstorms in Illinois, Report Champaign, Illinois, USA.

Idornigie E., Templeton M. R., Maksimovic C. & Sharifan S. 2010 The impact of variable hydraulic operation of water distribution networks on disinfection by-product concentrations. Urban Water Journal 7, 301-307 doi:10.1080/1573062X.2010.509438.

Israel G. D. 1992 Determining sample size, Report Fact Sheet PEOD-6, Program Evaluation and Organizational Development, Florida Cooperative Extension Service, University of Florida. PEOD-6. November,

Jansen T. 2013 Analyzing evolutionary algorithms: The computer science perspective. Springer Science & Business Media,

Jin T. 2019 Reliability engineering and services. John Wiley & Sons, Inc., Hoboken, NJ, USA

Jin Y. & Sendhoff B. 2003 Trade-off between Performance and Robustness: An Evolutionary Multiobjective Approach. In: Second International Conference on Evolutionary Multi-Criterion Optimization, EMO 2003, Faro, Portugal. Springer, pp 237-251

Jolly M. D., Lothes A. D., Sebastian Bryson L. & Ormsbee L. 2013 Research database of water distribution system models. Journal of Water Resources Planning and Management 140, 410-416 doi:10.1061/(ASCE)WR.1943-5452.0000352.

Kang D. & Lansey K. 2009 Real-time optimal valve operation and booster disinfection for water quality in water distribution systems. Journal of Water Resources Planning and Management 136, 463-473 doi:10.1061/(ASCE)WR.1943-5452.0000056.

Kang D. & Lansey K. 2012 Scenario-based robust optimization of regional water and wastewater infrastructure. Journal of Water Resources Planning and Management 139, 325-338

Kapelan Z. S., Savic D. A. & Walters G. A. 2005 Multiobjective design of water distribution systems under uncertainty. Water Resources Research 41, 1-15 doi:10.1029/2004WR003787.

Kapelan Z. S., Savic D. A., Walters G. A. & Babayan A. V. 2006 Risk- and robustness-based solutions to a multi-objective water distribution system rehabilitation problem under uncertainty. Water Science & Technology 53, 61-75 doi:10.2166/wst.2006.008.

Karmeli D., Gadish Y. & Meyers S. 1968 Design of optimal water distribution networks. Journal of the Pipeline Division 94, 1-10

Karovic O. & Mays L. W. 2014 Sewer system design using simulated annealing in Excel. Water Resources Management 28, 4551-4565

Kebede S. A. 2014 Optimal design of urban stormwater drainage system under uncertainity. Master's thesis, UNESCO-IHE Institute for Water Education, Delft, The Netherlands.

Keifer C. J. & Chu H. H. 1957 Synthetic storm pattern for drainage design. Journal of the hydraulics division 83, 1-25

Klein G., Moskowitz H. & Ravindran A. 1990 Interactive Multiobjective Optimization under Uncertainty. Institute for Operations Research and the Managements Sciences, 58-75

Klotz D., Strafaci A. & Totz C. 2007 Stormwater Conveyance Modeling and Design. Bentley Institute Press,

Kollat J. B. & Reed P. M. 2006 Comparing state-of-the-art evolutionary multi-objective algorithms for long-term groundwater monitoring design. Advances in Water Resources 29, 792-807

Kukkonen S. & Deb K. 2006 A Fast and Effective Method for Pruning of Non-Dominated Solutions in Many-Objective Problems. In: 9th International Conference on Parallel Problem Solving from Nature (PPSN IX). Reykjavik, Iceland. Springer, pp 553-562

Kuzmin V. A. 2009 Algorithms of Automatic Calibration of Multi-parameter Models Used in Operational Systems of Flash Flood Forecasting. Russian Meteorology and Hydrology 34, 92-104

Loucks D. P. & Van Beek E. 2017 Water resource systems planning and management: An introduction to methods, models, and applications. Springer,

Machell J. & Boxall J. 2014 Modeling and field work to investigate the relationship between age and quality of tap water. Journal of Water Resources Planning and Management 140, 04014020 doi:10.1061/(ASCE)WR.1943-5452.0000383.

Machell J., Boxall J., Saul A. & Bramley D. 2009 Improved representation of water age in distribution networks to inform water quality. Journal of Water Resources Planning and Management 135, 382-391 doi:10.1061/(ASCE)0733-9496(2009)135:5(382).

Maharjan M., Pathirana A., Gersonius B. & Vairavamoorthy K. 2008 Staged cost optimization of urban storm drainage systems based on hydraulic performance in a changing environment. Hydrology and Earth System Sciences Discussions 5, 1479-1509

Maier H. R., Kapelan Z., Kasprzyk J., Kollat J., Matott L. S., Cunha M. C., Dandy G. C., Gibbs M. S., Keedwell E., Marchi A., Ostfeld A., Savic D., Solomatine D. P., Vrugt J. A., Zecchin A. C., Minsker B. S., Barbour E. J., Kuczera G., Pasha F., Castelletti A., Giuliani M. & Reed P. M. 2014 Evolutionary algorithms and other metaheuristics in water resources: Current status, research challenges and future directions. Environmental Modelling & Software 62, 271-299

Mala-Jetmarova H., Sultanova N. & Savic D. 2017 Lost in optimisation of water distribution systems? A literature review of system operation. Environmental Modelling & Software 93, 209-254 doi:10.1016/j.envsoft.2017.02.009.

Marchi A., Dandy G. C. & Maier H. R. 2016 Integrated approach for optimizing the design of aquifer storage and recovery stormwater harvesting schemes accounting for externalities and climate change. Journal of Water Resources Planning and Management 142, 04016002

Marchi M., Rizzian L., Rigoni E., Russo R. & Clarich A. 2014 Combining robustness and reliability with polynomial chaos techniques in multiobjective optimization problems: use of percentiles. In: Proceedings of the 9th International Conference on Structural Dynamics, EURODYN 2014, Porto, Portugal. pp 2981-2988

Marsalek J. & Watt W. 1984 Design storms for urban drainage design. Canadian Journal of Civil Engineering 11, 574-584

Martinez-Cano C., Toloh B., Sanchez Torres A., Vojinovic Z. & Brdjanovic D. 2014 Flood resilience assessment in urban drainage systems through multi-objective optimisation. Paper presented at the 11th International Conference on Hydroinformatics, HIC 2014, New York City, USA,

Masters S., Parks J., Atassi A. & Edwards M. A. 2015 Distribution system water age can create premise plumbing corrosion hotspots. Environmental monitoring and assessment 187, 559 doi:10.1007/s10661-015-4747-4.

McPhail C., Maier H., Kwakkel J., Giuliani M., Castelletti A. & Westra S. 2018 Robustness metrics: How are they calculated, when should they be used and why do they give different results? Earth's Future 6, 169-191 doi:10.1002/2017EF000649.

Menapace A., Avesani D., Righetti M., Bellin A. & Pisaturo G. 2018 Uniformly distributed demand EPANET extension. Water resources management 32, 2165-2180

Milly P. C., Betancourt J., Falkenmark M., Hirsch R. M., Kundzewicz Z. W., Lettenmaier D. P. & Stouffer R. J. 2008 Stationarity is dead: Whither water management? Science 319, 573-574 doi:10.1126/science.1151915.

Morris R. D., Audet A.-M., Angelillo I. F., Chalmers T. C. & Mosteller F. 1992 Chlorination, chlorination by-products, and cancer: a meta-analysis. American journal of public health 82, 955-963

Mortazavi-Naeini M., Kuczera G., Kiem A. S., Cui L., Henley B., Berghout B. & Turner E. 2015 Robust optimization to secure urban bulk water supply against extreme drought and uncertain climate change. Environmental Modelling & Software 69, 437-451

Nicklow J., Reed P., Savic D., Dessalegne T., Harrell L., Chan-Hilton A., Karamouz M., Minsker B., Ostfeld A., Singh A. & Zechman E. 2009 State of the art for genetic algorithms and beyond in water resources planning and management. Journal of Water Resources Planning and Management 136, 412-432

Novak P., Guinot V., Jeffrey A. & Reeve D. E. 2010 Hydraulic modelling–an Introduction. Spon Prees, New York, USA,

Odan F. K., Ribeiro Reis L. F. & Kapelan Z. 2015 Real-time multiobjective optimization of operation of water supply systems. Journal of Water Resources Planning and Management 141, 04015011

Ong Y. S. N., P. B. & Lum K. Y. 2006 Max-Min Surrogate-Assisted Evolutionary Algorithm for Robust Design. Transactions en Evolutionary Computation,

Organization W. H. 2004 Guidelines for drinking-water quality: recommendations vol 1. World Health Organization,

Ostfeld A. & Salomons E. 2006 Conjunctive optimal scheduling of pumping and booster chlorine injections in water distribution systems. Engineering Optimization 38, 337-352 doi:10.1080/03052150500478007.

Pappenberger F. & Beven J. 2006 Ignorance is bliss: Or seven reasons not to use uncertainty analysis. Water Resources Research 42, 1-8

Pasha M. & Lansey K. 2005 Analysis of uncertainty on water distribution hydraulics and water quality. Paper presented at the World Water and Environmental Resources Congress 2005, Anchorage, Alaska, United States,

Patra K. C. 2008 Hydrology and water resources engineering. 2nd edn. Alpha Science International, Oxford, U.K.

Petrone G., Iaccarino G. & Quagliarella D. 2011a Robustness criteria in optimization under uncertainty. In: Evolutionary and Deterministic Methods for Design, Optimization and Control (Eurogen 2011), Capua, Italia. Springer,

Petrone G., Nicola C. d., Quagliarella D., Witteveen J., Axerio-Cilies J. & Iaccarino G. 2011b Wind turbine optimization under uncertainty with high performance computing. In: 29th AIAA Applied Aerodynamics Conference, Honolulu,Hawaii. American Institute of Aeronautics and Astronautics,

Prasad T. D. & Walters G. A. 2006 Minimizing residence times by rerouting flows to improve water quality in distribution networks. Engineering Optimization 38, 923-939 doi:10.1080/03052150600833036.

Quintiliani C., Alfonso L., Di Cristo C., Leopardi A. & de Marinis G. 2017 Exploring the use of operational interventions in water distribution systems to reduce the formation of TTHMs. Procedia Engineering 186, 475-482 doi:10.1016/j.proeng.2017.03.258.

Quintiliani C., Marquez-Calvo O. O., Alfonso L., Di Cristo C., Leopardi A., Solomatine D. P. & de Marinis G. 2019 Multi-objective valve management optimization formulations for water quality enhancement in WDNs. Journal of Water Resources Planning and Management 145, 04019061 doi:10.1061/(ASCE)WR.1943-5452.0001133.

Reed P. M., Hadka D., Herman J. D., Kasprzyk J. R. & Kollat J. B. 2013 Evolutionary multiobjective optimization in water resources: The past, present, and future. Advances in Water Resources 51, 438–456

Reyes-Sierra M. & Coello-Coello C. A. 2005 Improving PSO-based Multi-Objective Optimization using Crowding, Mutation and epsilon-Dominance. In: Evolutionary Multi-Criterion Optimization: Third International Conference, EMO 2005, Guanajuato, Mexico. Springer, pp 505-519

Roach T., Kapelan Z., Ledbetter R. & Ledbetter M. 2016 Comparison of robust optimization and info-gap methods for water resource management under deep uncertainty. Journal of Water Resources Planning and Management 142, 04016028

Rossman L. A. 1999 The EPANET programmer's toolkit for analysis of water distribution systems. Paper presented at the 29th Annual Water Resources Planning and Management Conference, Tempe, Arizona, United States,

Rossman L. A. 2010 Storm Water Management Model Version 5.0. Water Resources Division of the U.S. Environmental Protection Agency and Camp Dresser & McKee Inc., Cincinnati, Ohio, United States of America

Samora I., Franca M. J., Schleiss A. J. & Ramos H. M. 2015 Optimal location of micro-turbines in water supply network. In: 36th IAHR World Congress, The Hague, The Netherlands. vol CONF.

Samora I., Franca M. J., Schleiss A. J. & Ramos H. M. 2016 Simulated annealing in optimization of energy production in a water supply network. Water Resources Management 30, 1533-1547 doi:10.1007/s11269-016-1238-5.

Savic D. 2005 Coping with Risk and Uncertainty in Urban Water Infrastructure Rehabilitation Planning. In: Acqua e Città - I Convegno Nazionale di Idraulica Urbana, Sant'Agnello, Naples, Italy.

Savic D. 2006 Robust design and management of water systems: How to cope with risk and uncertainty? In: Integrated Urban Water Resources Management, P. H., T. K., J. M., I. M. (eds). NATO Security through Science Series. Springer, Dordrecht, pp 91-100. doi:10.1007/1-4020-4685-5_10.

Schaake J. C. & Lai F. H. 1969 Linear programming and dynamic programming application to water distribution network design, Report 116, Massachusetts Institute of Technology, Cambridge, Mass.

Schal S., Bryson L. S. & Ormsbee L. E. 2016 A simplified procedure for sensor placement guidance for small utilities. International Journal of Critical Infrastructures 12, 195-212

Sebti A., Bennis S. & Fuamba M. 2016 Optimization of the restructuring cost of an urban drainage network. Urban Water Journal 13, 119-132

Serrano S. E. 2011 Engineering Uncertainty and Risk Analysis, Second Edition: A Balanced Approach to Probability, Statistics, Stochastic Models, and Stochastic Differential Equations. ISBN 0965564312. HydroScience, Incorporated, Ambler, PA, USA

Seyoum A. G. & Tanyimboh T. T. 2017 Integration of hydraulic and water quality modelling in distribution networks: EPANET-PMX. Water resources management 31, 4485-4503

Shapiro A., Dentcheva D. & Ruszczynski A. 2009 Lectures in stochastic programming: Modeling and theory. SIAM-Society for Industrial and Applied Mathematics,

Shokoohi M., Tabesh M., Nazif S. & Dini M. 2017 Water quality based multi-objective optimal design of water distribution systems. Water Resources Management 31, 93-108 doi:10.1007/s11269-016-1512-6.

Solomatine D. P. 2012 Robust Optimization and Probabilistic Analysis of Robustness (ROPAR). http://www.un-ihe.org/hi/sol/papers/ROPAR.pdf. (accessed 15 January 2013).

Steele J. C., Mahoney K., Karovic O. & Mays L. W. 2016 Heuristic optimization model for the optimal layout and pipe design of sewer systems. Water Resources Management 30, 1605-1620

Stijnen J. W., Kanning W., Jonkman S. N. & Kok M. 2014 The technical and financial sustainability of the Dutch. Flood Risk Management 7, 3-15

Sullivan T. J. 2015 Introduction to uncertainty quantification vol 63. Springer,

Tiwari S., Fadel G. & Deb K. 2011 AMGA2: improving the performance of the archive-based micro-genetic algorithm for multi-objective optimization. Engineering optimization 43, 377-401

Toloh B. 2014 Assessment of resilience using multi-objective optimization for urban drainage rehabilitation measures. Master's thesis, UNESCO-IHE Institute for Water Education, Delft, The Netherlands.

Tricarico C., De Marinis G., Gargano R. & Leopardi A. 2007 Peak residential water demand. Proceedings of the Institution of Civil Engineers-Water Management 160, 115-121 doi:10.1680/wama.2007.160.2.115.

Tsimopoulou V., Kok M. & Vrijling J. K. 2015 Economic optimization of flood prevention systems in the Netherlands. Mitigation and Adaptation Strategies for Global Change,

Ulanicki B. & Kennedy P. R. 1994 An optimization technique for water network operations and design. In: 33rd IEEE Conference on Decision and Control, Lake Buena Vista, FL, USA. IEEE, pp 4114-4115. doi:10.1109/CDC.1994.411590.

van Dijk E., van der Meulen J., Kluck J. & Straatman J. 2014 Comparing modelling techniques for analysing urban pluvial flooding. Water science and technology 69, 305-311

Velez Quintero C. A. 2012 Optimization of urban wastewater systems using model based design and control. Doctoral dissertation, TU Delft, Delft University of Technology, Delft, Netherlands.

Vojinovic Z., Sahlu S., Torres A. S., Seyoum S. D., Anvarifar F., Matungulu H., Barreto W., Savic D. & Kapelan Z. 2014 Multi-objective rehabilitation of urban drainage systems under uncertainties. Journal of Hydroinformatics 16, 1044-1061

Vrugt J. A. & Robinson B. A. 2007 Improved evolutionary optimization from genetically adaptive multimethod search. Proc Natl Acad Sci USA 104, 708-711

Walker W. E., Haasnoot M. & Kwakkel J. H. 2013 Adapt or perish: a review of planning approaches for adaptation under deep uncertainty. Sustainability 5, 955-979

Watson A. A. & Kasprzyk J. R. 2017 Incorporating deeply uncertain factors into the many objective search process. Environmental Modelling & Software 89, 159-171

Weinberg H. S., Krasner S. W., Richardson S. D. & Thruston Jr A. D. 2002 The occurrence of disinfection by-products (DBPs) of health concern in drinking water: results of a nationwide DBP occurrence study, Report EPA/600/R-02/068 (NTIS PB2003-106823), U.S. Environmental Protection Agency, National Exposure Research Laboratory, Athens, GA, USA.

Willems P. 2008 Quantification and relative comparison of different types of uncertainties in sewer water quality modeling. Water research 42, 3539-3551 doi:10.1016/j.watres.2008.05.006.

Witteveen J. A. S. & Iaccarino G. 2010 Simplex Elements Stochastic Collocation for Uncertainty Propagation in Robust Design Optimization. In: 48th AIAA Aerospace Sciences Meeting Including the New Horizons Forum and Aerospace Exposition, Orlando, Florida, USA.

Yazdi J., Lee E. H. & Kim J. H. 2014 Stochastic multiobjective optimization model for urban drainage network rehabilitation. Journal of Water Resources Planning and Management 141, 04014091

Yazdi J., Yoo D. G. & Kim J. H. 2016 Comparative study of multi-objective evolutionary algorithms for hydraulic rehabilitation of urban drainage networks. Urban Water Journal,

Zeferino J. A., Cunha M. C. & Antunes A. P. 2012 Robust optimization approach to regional wastewater system planning. Journal of environmental management 109, 113-122

Zhang Q., Liu W. & Li H. 2009 The Performance of a New Version of MOEA/D on CEC09 Unconstrained MOP Test Instances, Report Colchester, Essex, UK.

Zimmermann H.-J. 2000 An application-oriented view of modeling uncertainty. European Journal of operational research 122, 190-198

Zitzler E., Deb K. & Thiele L. 2000 Comparison of Multiobjective Evolutionary Algorithms: Empirical Results. Evolutionary Computation 8, 173-195

Zitzler E. & Kunzli S. 2004 Indicator-Based Selection in Multiobjective Search. In: Proc. Parallel Problem Solving from Nature - PPSN VIII: 8th International Conference, Birmingham, UK. Springer, pp 832-842

**Netherlands Research School for the
Socio-Economic and Natural Sciences of the Environment**

DIPLOMA

For specialised PhD training

The Netherlands Research School for the
Socio-Economic and Natural Sciences of the Environment
(SENSE) declares that

Oscar Osvaldo
Marquez Calvo

born on 31 May 1967 in Oaxaca, Mexico

has successfully fulfilled all requirements of the
Educational Programme of SENSE.

Delft, 15 January 2020

The Chairman of the SENSE board

Prof. dr. Martin Wassen

the SENSE Director of Education

Dr. Ad van Dommelen

KONINKLIJKE NEDERLANDSE
AKADEMIE VAN WETENSCHAPPEN

The SENSE Research School declares that Oscar Osvaldo Marquez Calvo has successfully
fulfilled all requirements of the Educational PhD Programme of SENSE with a
work load of 35.7 EC, including the following activities:

SENSE PhD Courses

o Research in context activity: 'Active contribution to development and implementation of
Knowledge Base platform as online tool: http://www.kulturisk.eu/knowledge-base
(2013)

Other PhD and Advanced MSc Courses

o Discover AI workshop with Microsoft Azure, TU Delft (2019)
o Hydrology and hydraulics, IHE Delft (2013)
o Data driven modelling and real-time control of water systems, IHE Delft (2013)
o Hydroinformatics for decision support, IHE Delft (2013)
o 1D and 2D water systems and watershed modelling with US EPA SWMM5 and PCSWMM
Europe, IHE Delft (2018)
o Introduction to Google Earth engine for hydrological sciences, European Geosciences
Union (2016)
o Introductory time-series analysis: how to apply and interpret the Fast Fourier Transform,
European Geosciences Union (2016)
o Advanced academic writing course for PhD fellows, IHE Delft (2018)
o Post-PhD Career Planning Workshop, IHE Delft (2017)

Management and Didactic Skills Training

o Supervising MSc student with thesis entitled 'Uncertainty-Aware multiobjective
optimization of urban drainage systems using surrogate models' (2013)

Oral and Poster Presentations

o *Towards robust optimal design of storm water systems.* EGU General Assembly, 13-17
April 2015, Vienna, Austria
o *Comparative analysis of ROPAR, a method to find robust optimum solutions to problems
with multiple objectives.* EGU General Assembly, 8-12 April 2019, Vienna, Austria

SENSE Coordinator PhD Education

Dr. ir. Peter Vermeulen

T - #0104 - 071024 - C184 - 240/170/10 - PB - 9780367460433 - Gloss Lamination